APP MONETIZATION MASTERY

Ulas Can Cengiz

Disclaimer

The information provided in this book is intended for informational and educational purposes only and should not be construed as legal, financial, or professional advice. The author has made every effort to ensure the accuracy and completeness of the information provided, but they cannot guarantee that it is without errors or omissions.

While the author has drawn from their own experiences as a software developer and entrepreneur, they are not a legal expert. The legal and regulatory information provided in this book is not exhaustive and may not be applicable to all situations or jurisdictions. Laws, regulations, and best practices are subject to change, and the specific requirements for your app may vary based on factors such as its nature, target audience, and location.

Readers are encouraged to consult with legal, financial, and business professionals to obtain advice tailored to their unique circumstances before implementing any strategies or taking any actions discussed in this book. The author disclaims any liability for any loss or damage that may arise from the use of, or inability to use, the information contained in this book.

By reading this book, you acknowledge that you are solely responsible for the decisions you make regarding your app's development, monetization, and compliance with applicable laws and regulations. Neither the author nor the publisher assumes any responsibility for any actions taken, or not taken, based on the information provided in this book.

Foreword

As a software developer and entrepreneur with over 15 years of experience in the industry, I have seen firsthand the rapid growth and immense potential that lies within the mobile app market. Today, apps have become an integral part of our daily lives, offering solutions to a wide range of problems and fulfilling diverse needs. With this expansion comes the opportunity for developers, entrepreneurs, and businesses to create profitable ventures through effective app monetization.

"App Monetization Mastery: Strategies for Turning Your App Idea into a Profitable Business" is the culmination of my experience and insights gained throughout my career. My goal in writing this book is to provide you with a comprehensive guide that covers the essential aspects of app development and monetization, while also offering valuable advice on more advanced strategies for achieving success.

In addition to providing you with the knowledge contained in this book, I am also proud to offer consulting services tailored to the unique needs of your app business. By working directly with me, you can benefit from personalized guidance and support, helping you to navigate the complex world of app monetization with greater confidence and success.

In this book, you will find a wealth of information on various monetization models, including in-app advertising, in-app purchases, and subscription-based revenue streams. Additionally, the book explores critical aspects of app growth and expansion, such as internationalization, cross-platform development, and strategic partnerships.

One of the core principles emphasized in this book is the importance of data-driven decision-making when it comes to app monetization. By leveraging analytics to understand user behavior and preferences, you can optimize your app's revenue generation and drive growth.

As you read through the pages of this book, remember that successful app monetization is a combination of art and science. It requires a deep understanding of your target audience, a willingness to experiment with different strategies, and the ability to adapt to ever-changing market conditions. I am confident that "App Monetization Mastery: Strategies for Turning Your App Idea into a Profitable Business" will serve as a valuable resource on your journey towards app monetization success.

Whether you are an experienced developer looking to refine your monetization skills, an aspiring entrepreneur seeking to turn your app idea into a profitable business, or a business owner in search of guidance for maximizing your app's revenue potential, this book and my consulting services will prove to be indispensable companions on your journey. I wish you the best of luck in your app monetization endeavors and hope that the insights shared in this book, along with my personalized support, inspire you to unlock your app's full potential.

Sincerely,

Ulas Can Cengiz

CHAPTER ONE
Introduction

Why App Monetization Matters

The rapid expansion of smartphone usage and the increasing adoption of mobile applications have turned app development into a lucrative industry. As a software developer and entrepreneur with 15+ years of experience, I have witnessed the incredible potential that well-executed app monetization strategies can offer. Let's delve into the reasons why app monetization matters and how it can contribute to the success of your app and your business.

The Importance of Revenue Generation

The most fundamental reason why app monetization matters is that it allows you to generate revenue from your app. Developing, maintaining, and marketing a mobile app require substantial time, effort, and resources. To justify these investments, you need a well-thought-out and effective monetization strategy that will help you generate a steady stream of income. This revenue not only covers the costs associated with app development but also allows you to invest in future updates, improvements, and new projects.

Sustaining Long-Term Growth

A successful monetization strategy is crucial for sustaining the long-term growth of your app and your business. As your app generates more revenue, you can reinvest those profits into marketing efforts, reaching a wider audience, and expanding your user base. This, in turn, can lead to even greater revenue generation and further growth opportunities. By continuously iterating and optimizing your monetization strategy, you can create a virtuous cycle that drives the success of your app and your business.

User Acquisition and Retention

Another reason why app monetization matters is its influence on user acquisition and retention. An effective monetization strategy can actually enhance the user experience by offering users valuable content, features, or services in exchange for their financial support. By providing users with compelling reasons to spend money on your app, you can foster a sense of loyalty and encourage them to continue using your app over time. Additionally, by understanding how users engage with and respond to your monetization efforts, you can identify opportunities to improve the user experience and boost retention rates.

Competitive Advantage

In today's highly competitive app market, having a strong monetization strategy is a key differentiator. Apps that effectively generate revenue can invest more in marketing, research, and development, giving them a competitive edge over rivals. Moreover, a well-executed monetization plan sends a signal to potential investors, partners, and users that your app is financially viable and has a clear path to profitability. This can help attract interest, build trust, and establish your app as a leader in its category.

Fostering Innovation and Creativity

A robust monetization strategy can also foster innovation and creativity within your app development team. By generating a steady revenue stream, you can allocate more resources towards research and development, enabling your team to experiment with new features, technologies, and strategies. This can lead to the creation of unique and engaging app experiences that not only delight users but also drive revenue and growth.

Supporting Your Business and Team

Finally, app monetization matters because it supports the financial well-being of your business and team. As an entrepreneur, your primary goal is to build a successful and profitable business that supports not only yourself but also your employees. By implementing a successful app monetization strategy, you can generate the revenue necessary to pay your team, invest in their professional development, and create a positive work environment that fosters collaboration and creativity.

In conclusion, app monetization is an essential aspect of turning your app idea into a profitable business. It enables you to generate revenue, sustain long-term growth, acquire and retain users, gain a competitive advantage, foster innovation, and support your business and team. As we progress through this book, we will explore various strategies and best practices that will help you master app monetization and achieve success in the ever-evolving world of mobile applications.

Setting the Stage for Success

To succeed in the competitive world of app development and monetization, you need a solid foundation upon which to build your strategies. In this subchapter, we will discuss various factors to consider and steps to take to set the stage for success in your app monetization journey.

Define Your App's Unique Value Proposition

Before you can effectively monetize your app, it is crucial to identify and articulate its unique value proposition (UVP). Your UVP is what sets your app apart from competitors and makes it valuable to users. Start by asking yourself the following questions:

- What problem does my app solve?
- Who is my target audience?
- What features or benefits does my app offer that competitors don't?
- Why should users choose my app over similar options?

By defining your app's UVP, you can better understand how to position your app in the market and develop a monetization strategy that aligns with your app's core strengths.

Conduct Thorough Market Research

Understanding your app's target market and its competitors is essential for developing a successful monetization strategy. Conduct thorough market research to gain insights into user preferences, needs, and behaviors. This information will help you identify gaps in the market, refine your app's features and functionality, and create a compelling pricing strategy that appeals to your target audience.

Additionally, researching your competitors' monetization strategies can provide valuable insights into what works and what doesn't. By analyzing successful apps in your niche, you can identify best practices and potential opportunities for differentiation.

Design Your App for Monetization from the Start

To maximize your app's monetization potential, it is crucial to design your app with monetization in mind from the start. This means integrating monetization features seamlessly into the user experience and ensuring that they align with your app's UVP.

For example, if your app offers a freemium model, make sure the premium features are genuinely valuable and enticing to users. If you opt for in-app advertising, consider ad formats and placements that will not detract from the user experience. By thinking about monetization during the design and development stages, you can create an app that generates revenue without sacrificing user satisfaction.

Develop a Data-Driven Monetization Strategy

Data-driven decision-making is essential for creating a successful app monetization strategy. By leveraging data and analytics, you can gain valuable insights into user behavior and preferences, enabling you to optimize your monetization efforts.

Invest in robust analytics tools that can track key performance indicators (KPIs) related to monetization, such as average revenue per user (ARPU), conversion rates, and user lifetime value (LTV). Analyze this data regularly to identify trends, opportunities for improvement, and areas where your monetization efforts may be falling short. Use this information to make informed decisions about pricing, offers, promotions, and other monetization tactics.

Test and Optimize Your Monetization Strategy

No monetization strategy is perfect from the outset. It is essential to continuously test and optimize your monetization efforts to maximize their effectiveness. This can involve A/B testing different pricing structures, ad formats, and in-app purchase offers to determine which resonate best with your users.

Be prepared to iterate and adapt your monetization strategy based on user feedback and performance data. By staying agile and responsive to changes in user behavior and market conditions, you can ensure that your monetization efforts remain effective and profitable over time.

Build a Strong Marketing and User Acquisition Plan

An effective monetization strategy is only as successful as your ability to attract and retain users. Therefore, it is crucial to develop a strong marketing and user acquisition plan that can help you build awareness, drive app downloads, and engage users.

Your marketing plan should include a mix of strategies such as app store optimization (ASO), paid advertising, content marketing, social media, influencer partnerships, and public relations. By leveraging multiple channels, you can increase the reach and visibility of your app, attracting a larger and more diverse user base.

Invest time and resources into refining your app's onboarding process and user experience, as these factors play a critical role in user retention. The more engaged and satisfied your users are, the more likely they are to spend money on your app and recommend it to others.

Foster a Strong Company Culture and Team

The success of your app monetization efforts is largely dependent on the talent and dedication of your team. By fostering a strong company culture and investing in the professional development of your employees, you can create an environment that encourages innovation, collaboration, and success.

Encourage open communication and feedback, and provide opportunities for your team members to grow and learn. By empowering your employees to take ownership of their work and contribute to the app's success, you can foster a sense of pride and commitment that will ultimately drive better results.

Stay Informed and Adapt to Industry Trends

The mobile app industry is constantly evolving, with new technologies, platforms, and monetization models emerging regularly. To maintain a successful app monetization strategy, it is essential to stay informed about industry trends and be prepared to adapt your approach accordingly.

Regularly attend industry conferences, read reputable blogs and publications, and participate in online forums to stay up-to-date on the latest developments in app monetization. By staying informed and remaining flexible, you can ensure that your monetization strategy remains relevant and effective in an ever-changing landscape.

Measure Success and Continuously Improve

Finally, it is crucial to establish clear objectives and KPIs for your app monetization efforts and track your progress against these benchmarks. By regularly monitoring your performance, you can identify areas of strength and opportunities for improvement.

Use this information to refine your monetization strategy, set new goals, and iterate on your approach. By continuously improving your app monetization efforts and striving for excellence, you can maximize your app's revenue potential and achieve long-term success.

In conclusion, setting the stage for success in app monetization involves defining your app's UVP, conducting thorough market research, designing your app for monetization, leveraging data and analytics, testing and optimizing your strategy, building a strong marketing and user acquisition plan, fostering a strong company culture and team, staying informed about industry trends, and measuring success to continuously improve. By taking these steps, you can create a strong foundation upon which to build your app monetization mastery and turn your app idea into a profitable business.

CHAPTER TWO

Laying the Foundation: Pre-Launch Monetization Strategies

Market Research and Analysis

Identifying Your Target Audience

One of the key factors for successfully monetizing your app is to have a clear understanding of your target audience. By identifying your target audience, you can tailor your app's features, user experience, and monetization strategies to resonate with their preferences, needs, and behaviors. In this section, we will explore the process of identifying your target audience and understanding their unique characteristics.

To begin, start by defining the demographics of your target audience, which includes factors such as age, gender, location, income, and education level. This information will provide you with a foundation for understanding who is most likely to use and benefit from your app.

Next, dive deeper into the psychographics of your target audience. This includes aspects such as their interests, hobbies, values, and lifestyle choices. By examining the psychographics of your audience, you can develop a more nuanced understanding of their motivations and how your app can cater to their needs.

* * *

Once you have gathered demographic and psychographic information about your target audience, create user personas that represent various segments of your audience. User personas are fictional representations of your ideal users, based on the data you have collected. They help you visualize and empathize with your users, enabling you to make more informed decisions about your app's design, functionality, and monetization strategies.

Finally, validate your target audience assumptions by conducting surveys, interviews, and focus groups with potential users. This will help you gain firsthand insights into their preferences, pain points, and expectations, allowing you to refine your understanding of your target audience and make necessary adjustments to your app and monetization strategies.

Analyzing Competitor Strategies

In the competitive world of app development, it is crucial to analyze the strategies of your competitors to identify best practices and potential opportunities for differentiation. By understanding what works and what doesn't for other apps in your niche, you can develop a more effective monetization strategy that sets your app apart from the competition. In this section, we will discuss the process of analyzing competitor strategies and using this information to inform your own monetization efforts.

Begin by identifying the top competitors in your app's niche or category. These competitors may include both direct competitors, who offer similar products or services, and indirect competitors, who cater to the same target audience but with different offerings. Create a list of these competitors and gather as much information as possible about their monetization strategies, including their pricing models, in-app purchases, advertising efforts, and partnerships.

Once you have collected information on your competitors, analyze their strategies in detail. Look for patterns and trends in their monetization efforts, and identify any unique or innovative approaches that you can learn from or adapt to your own app. Consider the following questions as you analyze your competitors' strategies:

* * *

- What monetization models are they using, and which ones seem to be the most effective?
- How do they price their in-app purchases or subscriptions, and what factors seem to influence their pricing decisions?
- What types of ads or sponsored content do they feature in their app, and how do they balance monetization with user experience?
- Are there any partnerships or collaborations that contribute to their monetization efforts, and if so, how do these partnerships work?

After analyzing your competitors' strategies, evaluate the strengths and weaknesses of their approaches. This will help you identify opportunities for your own app to excel and outperform your competitors. For example, you might discover that a competitor's pricing model is too complex or that their in-app ads are intrusive and negatively affect the user experience. Use these insights to refine your own monetization strategy, ensuring that it addresses the shortcomings of your competitors while capitalizing on their strengths.

In conclusion, conducting thorough market research and analysis is essential for laying the foundation for successful app monetization. By identifying your target audience and understanding their unique characteristics, you can tailor your app and monetization strategies to resonate with their preferences and needs. Meanwhile, analyzing competitor strategies enables you to learn from their successes and failures, identify opportunities for differentiation, and develop a more effective monetization strategy that sets your app apart from the competition. By investing time and effort into understanding your target audience and the competitive landscape, you can create a strong foundation for your app's monetization efforts and ultimately achieve long-term success.

Now that we have discussed the importance of market research and analysis, let's move on to other essential pre-launch monetization strategies that will help you lay the foundation for a successful and profitable app.

In the next sections, we will explore additional pre-launch strategies, such as developing a data-driven pricing model, designing your app with monetization in mind, creating an effective marketing plan, and building a strong team to support your app's development and monetization efforts. By incorporating these strategies into your pre-launch planning process, you can set the stage for success in your app monetization journey and ensure that your app is well-positioned to generate revenue and achieve long-term profitability.

As you progress through the app development and monetization process, remember that your market research and competitor analysis should be ongoing activities. Regularly revisit your target audience assumptions and competitor strategies to ensure that your app remains relevant and competitive in the ever-evolving app marketplace. By staying informed and adapting your strategies as needed, you can maintain a successful app monetization strategy and continue to grow your app's revenue potential.

In summary, laying the foundation for successful app monetization involves understanding your target audience, analyzing competitor strategies, and incorporating these insights into your app's design, pricing, and marketing efforts. By taking the time to conduct thorough market research and analysis, you can create a strong foundation for your app's monetization efforts, setting the stage for success and long-term profitability. With this foundation in place, you can confidently move forward with the development and launch of your app, knowing that you have laid the groundwork for a successful and profitable app monetization strategy.

Designing Your App for Monetization

User Experience (UX) and Revenue

A well-designed user experience (UX) plays a critical role in the success of your app and its ability to generate revenue. An app that is easy to use, visually appealing, and engaging will not only attract and retain users but also encourage them to spend money within the app. In this section, we will discuss the importance of UX in app monetization and provide guidance on how to design your app to optimize both user satisfaction and revenue potential.

First and foremost, focus on creating a seamless and intuitive user experience. This includes designing an easy-to-navigate user interface (UI), optimizing app performance, and ensuring that your app's features and functionality align with the needs and preferences of your target audience. By prioritizing user satisfaction, you can increase user engagement, retention, and ultimately, the likelihood that users will spend money within your app.

To maximize revenue potential, incorporate monetization elements into your app's design from the very beginning. This means considering how your app's features, UI, and overall user experience can support various monetization strategies, such as in-app purchases (IAPs), subscriptions, or advertising. By designing your app with monetization in mind, you can create a more natural and compelling environment for users to spend money.

In-App Purchase (IAP) Integration

In-app purchases (IAPs) are a popular and effective monetization strategy that allows users to buy digital goods or services within your app. IAPs can include items such as virtual currency, extra lives, premium features, or additional content. To optimize the revenue potential of IAPs, it is crucial to design and integrate them in a way that enhances the user experience and encourages spending. In this section, we will discuss best practices for integrating IAPs into your app's design and maximizing their revenue potential.

1. Offer a Variety of IAPs: Provide a diverse range of IAP options that cater to different user preferences and spending habits. This can include a mix of consumable items (e.g., virtual currency or power-ups), non-consumable items (e.g., premium features or ad removal), and subscriptions (e.g., access to exclusive content). By offering a variety of IAPs, you can increase the likelihood that users will find something of value and be willing to spend money within your app.

2. Showcase IAPs Strategically: Position IAPs throughout your app in a way that is both visible and contextually relevant. This can include featuring IAPs on your app's home screen,

within game levels or content sections, and in strategic locations during the user journey (e.g., after completing a challenging level or reaching a milestone). By showcasing IAPs at the right time and place, you can increase their visibility and appeal to users.

3. Implement Seamless Payment Integration: Ensure that the payment process for IAPs is as simple and frictionless as possible. This includes providing multiple payment options (e.g., credit cards, mobile wallets, carrier billing), integrating with popular payment gateways, and streamlining the checkout flow. By making it easy for users to complete their purchases, you can increase the likelihood of conversion and reduce cart abandonment.

4. Create a Sense of Urgency: Encourage users to make IAPs by creating a sense of urgency or scarcity. This can include offering limited-time discounts, exclusive items, or special promotions. By creating a sense of urgency, you can incentivize users to make a purchase before the opportunity expires, driving revenue and user engagement.

5. Use Data to Optimize IAPs: Leverage data and analytics to understand how users interact with your IAPs and identify opportunities for improvement. This can include analyzing user behavior, purchase patterns, and price sensitivity, as well as conducting A/B testing to optimize the presentation, pricing, and promotion of your IAPs. By using data to inform your IAP strategy, you can ensure that your offerings are well-targeted, appealing, and effective at driving revenue.

6. Provide Value and Transparency: Ensure that your IAPs offer genuine value to users and that their benefits are clearly communicated. This includes providing detailed descriptions of what users can expect when they make a purchase, as well as showcasing reviews, ratings, or testimonials that demonstrate the value of your IAPs. By being transparent and providing value, you can build trust with users and encourage them to invest in your app.

7. Balance Monetization with User Experience: While it is important to maximize the revenue potential of your IAPs, it is crucial to strike a balance between monetization and user experience. Avoid being overly aggressive or intrusive with your IAP promotions, as this can alienate users and deter them

from spending money within your app. Instead, focus on creating a compelling and enjoyable user experience that naturally encourages users to engage with your IAPs and invest in your app.

By integrating IAPs into your app's design in a thoughtful and strategic manner, you can create a seamless and enjoyable user experience that drives revenue and supports long-term app success. Remember that the key to effective IAP monetization is to strike a balance between providing value, encouraging spending, and maintaining a high-quality user experience. By keeping these principles in mind, you can design your app for monetization and lay the foundation for a profitable and sustainable app business.

In conclusion, designing your app with monetization in mind is a critical aspect of laying the foundation for a successful app business. By prioritizing user experience and strategically integrating IAPs, you can create an engaging and revenue-generating app that appeals to your target audience and sets the stage for long-term success. As you continue to develop and refine your app monetization strategy, always keep the needs and preferences of your users at the forefront and remember that a high-quality user experience is ultimately the key to unlocking your app's revenue potential.

Building a Strong Business Model

Freemium vs. Premium

A crucial aspect of developing a successful app monetization strategy is selecting the right business model. Two popular business models for apps are freemium and premium. In this section, we will discuss the differences between these models, their pros and cons, and how to determine which is the best fit for your app.

Freemium: In the freemium model, users can download and use the core features of your app for free, while additional features, content, or services are available for purchase. This model can be highly effective in attracting a large user base, as the low barrier to entry allows users to try your app without any financial commitment. Once users are engaged with your app, they may be more willing to spend money on additional features or content that enhances their experience.

Pros:
- Low barrier to entry encourages user adoption.
- Can generate revenue through in-app purchases and/or advertising.
- Allows users to try before they buy, increasing the likelihood of conversion.

Cons:
- May require more marketing efforts to convert free users into paying customers.
- Potential for lower average revenue per user (ARPU) compared to premium apps.
- Can be challenging to strike a balance between free and paid features.

Premium: In the premium model, users pay a one-time fee upfront to download and use your app. This model can be advantageous for developers who have a well-established reputation, a highly differentiated app, or a niche target audience that is willing to pay for a premium experience. Premium apps typically have a higher ARPU compared to freemium apps, as users are required to pay upfront for access.

Pros:
- Higher ARPU compared to freemium apps.
- Simplified monetization strategy without the need for in-app purchases or advertising.
- Can create a sense of exclusivity and quality.

Cons:
- Higher barrier to entry may deter potential users.
- Requires a strong value proposition to justify the upfront cost.
- May have a smaller user base compared to freemium apps.

To determine which business model is the best fit for your app, consider factors such as your target audience, the competitive landscape, and your app's value proposition. If your app offers a unique or highly differentiated experience that users are willing to pay for, a premium model may be more suitable. Conversely, if your app has broad appeal and relies on user engagement to drive revenue, a freemium model may be more appropriate.

Subscription Models

Subscription models are becoming increasingly popular in the app market, as they provide developers with a steady stream of recurring revenue and allow users to access premium content and features on an ongoing basis. In this model, users pay a recurring fee (e.g., monthly, quarterly, or annually) to access your app's full suite of features and content.

Pros:
- Recurring revenue provides financial stability and predictability.
- Encourages user retention and long-term engagement.
- Can offer multiple subscription tiers to cater to different user preferences and budgets.

Cons:
- May require ongoing content updates or feature additions to justify the subscription cost.
- Can be challenging to acquire and retain subscribers.
- Subscription fatigue may deter potential users.

One-Time Purchases

One-time purchases are another monetization option that can be incorporated into your app's business model. This approach allows users to make a single payment for access to specific features, content, or services within your app. One-time purchases can be a viable option for apps that offer standalone products or services that do not require ongoing access or updates.

Pros:
- Simple and straightforward monetization strategy.
- Users only pay for the features or content they want.
- Can be combined with other monetization strategies, such as

freemium or subscription models.

Cons:
- May require more marketing efforts to drive sales of individual purchases.
- Less predictable revenue compared to subscription models.
- Can be challenging to maintain user engagement without ongoing updates or additions.

When choosing between freemium, premium, subscription, and one-time purchase models, it is essential to consider the unique characteristics of your app, your target audience, and your overall business goals. A combination of these models can also be utilized, depending on the structure and offerings of your app.

For example, you may decide to offer a freemium app with limited features and then provide users with the option to make one-time purchases for additional content or subscribe to a premium plan for ongoing access to all features. This hybrid approach allows you to cater to a broader range of user preferences and budgets while maximizing your app's revenue potential.

In conclusion, building a strong business model is a critical aspect of laying the foundation for a successful app monetization strategy. By carefully considering the pros and cons of each model and aligning your approach with your app's unique characteristics and target audience, you can create a monetization strategy that drives revenue and supports long-term app success.

As you continue to develop and refine your app's business model, be prepared to iterate and adapt your approach based on user feedback, market trends, and your app's performance metrics. By staying agile and responsive to the ever-evolving app market, you can position your app for sustained profitability and growth.

Laying the foundation for a successful app monetization strategy involves selecting the appropriate business model for your app, and this decision can have a significant impact on your app's overall revenue potential. Freemium, premium, subscription, and one-time purchase models each have their unique advantages and disadvantages, and it is essential to carefully consider these factors when designing your app's monetization strategy.

Ultimately, the success of your app's business model will depend on the value proposition you provide to your users, the competitive landscape of your market, and your ability to effectively communicate and deliver that value. By staying attuned to user feedback, market trends, and app performance metrics, you can refine and optimize your app's business model to maximize its revenue potential and long-term success.

CHAPTER THREE

Launch Strategies: Hitting the Ground Running

Effective App Store Optimization (ASO)

App Store Optimization (ASO) is a crucial component of any successful app launch strategy. ASO is the process of optimizing various aspects of your app's presence on app stores to improve its visibility, increase downloads, and ultimately boost your app's revenue. In this subchapter, we will delve into two key aspects of ASO: keyword research and usage, and optimizing your app icon and screenshots.

Keyword Research and Usage

Keywords play a vital role in ASO, as they are the primary method through which users find and discover apps in app stores. By conducting thorough keyword research and effectively utilizing these keywords in your app's metadata, you can significantly increase your app's visibility and attract more potential users.

1. Identify relevant keywords: Start by brainstorming a list of keywords that are relevant to your app, its features, and the problems it solves. Consider the language your target audience might use to search for your app or similar apps.
2. Conduct competitive analysis: Examine the keywords used by your competitors and identify any potential opportunities to differentiate your app or target less competitive keywords. This can help you gain a competitive edge in the crowded app

market.

3. Use keyword research tools: Leverage various keyword research tools, such as App Annie, Sensor Tower, or Mobile Action, to gain insights into keyword popularity, competition, and trends. These tools can help you identify high-potential keywords that have the best chance of driving organic traffic to your app.

4. Optimize your app's metadata: Incorporate your chosen keywords into your app's metadata, including the app title, subtitle, and description. Ensure that you strike a balance between keyword density and readability, as keyword stuffing can negatively impact your app's ranking and user perception.

5. Monitor and iterate: Regularly monitor your app's performance in terms of keyword rankings, downloads, and revenue. Be prepared to iterate on your keyword strategy based on your app's performance metrics and any changes in the competitive landscape.

App Icon and Screenshots

Your app icon and screenshots are the first visual elements users encounter when they discover your app on the app store. These visual assets play a crucial role in making a strong first impression and can significantly impact your app's conversion rate.

App Icon:

1. Design for impact: Your app icon should be eye-catching, memorable, and visually distinct from other apps in your category. A well-designed icon can help your app stand out in search results and encourage users to click through to your app's page.

2. Keep it simple: Avoid cluttering your app icon with too many details or text. Instead, focus on a single, recognizable element that represents your app's core functionality or brand. A simple, clean design is more likely to resonate with users and be easily remembered.

3. Test and iterate: Consider creating multiple versions of your app icon and testing them with your target audience or using app store A/B testing tools. By gathering feedback on different designs, you can optimize your app icon for maximum impact

and conversions.

Screenshots:
1. Showcase key features: Use your app screenshots to highlight the most important and engaging features of your app. This can help users quickly understand your app's value proposition and make an informed decision about whether to download your app.
2. Tell a visual story: Arrange your screenshots in a logical order that tells a visual story about your app's user experience. This can help users envision themselves using your app and encourage them to explore further.
3. Incorporate text overlays: Consider adding short, descriptive text overlays to your screenshots to provide additional context and explanation about your app's features. This can help users quickly grasp the benefits of your app and make it easier for them to understand how your app works.
4. Optimize for different devices: Ensure that your screenshots are optimized for different device screen sizes and resolutions. This can help ensure that your app looks polished and professional on a wide range of devices and improve the overall user experience.
5. A/B test your screenshots: Similar to your app icon, it can be beneficial to create multiple variations of your screenshots and test their performance using app store A/B testing tools. This can help you identify the most effective screenshot combinations and further optimize your app's conversion rate.

In conclusion, effective App Store Optimization (ASO) is a critical component of a successful app launch strategy. By conducting thorough keyword research and effectively incorporating those keywords into your app's metadata, you can increase your app's visibility and drive more organic traffic. Additionally, optimizing your app icon and screenshots can help you make a strong first impression on potential users, ultimately leading to higher conversion rates and increased revenue.

As you continue to refine your app's ASO strategy, be prepared to iterate and adapt based on your app's performance metrics, changes in the competitive landscape, and feedback from your target audience. By staying agile and responsive to the ever-evolving app market, you can position your app for sustained growth and profitability.

Remember, ASO is an ongoing process that requires constant attention and adjustments. As your app matures and the market evolves, you'll need to stay on top of new trends, competitor strategies, and user preferences to ensure that your app remains visible and attractive to potential users. By staying committed to ASO best practices and continuously optimizing your app's presence on app stores, you'll be well-positioned to achieve long-term success and turn your app idea into a profitable business.

PR and Influencer Marketing

In today's crowded app market, merely having a high-quality app is not enough to guarantee success. To stand out from the competition and attract users, it is essential to develop and execute a well-rounded marketing strategy. One powerful approach to building buzz and driving downloads is through public relations (PR) and influencer marketing. In this subchapter, we will explore two critical aspects of these strategies: crafting a compelling press release and identifying and engaging influencers.

Crafting a Compelling Press Release

A press release is a powerful tool that can help you generate media coverage and attract attention to your app. By crafting a well-written and engaging press release, you can pique the interest of journalists and bloggers, encouraging them to share your app with their audiences.

1.   Craft a newsworthy angle: To capture the attention of media professionals, your press release should present a newsworthy angle or story. Consider highlighting a unique feature, announcing a milestone, or sharing exciting news about your app's development or achievements.
2.   Write a captivating headline: The headline is the first thing journalists and readers will see, so make sure it is attention-grabbing and communicates the essence of your story. Keep it

concise and relevant to your target audience.

3.   Provide clear and concise information: In the body of your press release, provide all the necessary information about your app, including its features, benefits, and target audience. Be concise and focus on the most critical aspects of your app that will resonate with readers.

4.   Include quotes: Incorporate quotes from key stakeholders, such as the app's founder, developer, or satisfied users, to add credibility and human interest to your press release.

5.   Provide contact information: At the end of your press release, include contact information for media professionals to get in touch with you for further information or interview requests.

6.   Proofread and edit: Before distributing your press release, carefully proofread and edit it to ensure it is free of errors and conveys your message clearly and professionally.

7.   Distribute your press release: Once your press release is polished and ready for distribution, use PR distribution services or reach out to journalists and bloggers in your niche directly. Building relationships with media professionals can help increase the likelihood of coverage and ongoing support for your app.

Identifying and Engaging Influencers

Influencer marketing involves partnering with influential individuals, such as bloggers, YouTubers, or social media personalities, who can promote your app to their followers. By leveraging the trust and credibility these influencers have built with their audiences, you can drive awareness, downloads, and engagement for your app.

1.   Define your target audience: Before reaching out to influencers, it's essential to have a clear understanding of your target audience. This will help you identify influencers whose followers align with your app's user base and ensure that your marketing efforts resonate with the right people.

2.   Research potential influencers: Conduct thorough research to identify influencers who are relevant to your app's niche and have a strong following. Look for influencers with high engagement rates, as this indicates that their followers are more likely to take action upon their recommendations.

3. Evaluate influencers' fit: Before engaging with an influencer, assess their content, tone, and style to ensure they are a good fit for your app and brand. This can help create a more authentic and effective partnership.

4. Reach out with a tailored pitch: When approaching influencers, craft a personalized pitch that highlights the unique value of your app and explains how it would be relevant and valuable to their audience. Be clear about your expectations and what you can offer in return, such as free access to your app, monetary compensation, or promotional support.

5. Collaborate on content creation: Work closely with influencers to develop engaging and relevant content that promotes your app while resonating with their audience. This could include sponsored blog posts, social media posts, video reviews, or even co-branded campaigns. By collaborating on content creation, you can ensure that your app is showcased in a way that is both authentic to the influencer's style and appealing to their followers.

6. Track and analyze results: Once your influencer marketing campaign is underway, track its performance using relevant metrics such as impressions, engagement, click-throughs, and app downloads. This will help you evaluate the effectiveness of your influencer partnerships and make data-driven decisions about future collaborations.

7. Maintain relationships with influencers: Building long-term relationships with influencers can be highly beneficial for your app's ongoing success. Stay in touch with influencers, keep them updated on your app's progress, and explore opportunities for future collaborations. By nurturing these relationships, you can create a network of advocates who can help amplify your app's reach and visibility.

In summary, leveraging PR and influencer marketing strategies can be instrumental in promoting your app and generating buzz around its launch. By crafting a compelling press release and engaging with relevant influencers, you can tap into their established audiences and drive awareness, downloads, and engagement for your app. As you embark on your PR and influencer marketing journey, be prepared to iterate and adapt your approach based on your app's performance, industry trends, and feedback from your target audience. By staying agile and responsive to the ever-evolving app market, you can position your app for sustained growth and profitability, turning your app idea into a successful business venture.

User Acquisition Strategies

Driving app downloads and user acquisition is critical for the success of your app business. A well-executed user acquisition strategy can help you attract new users, increase engagement, and ultimately generate more revenue. In this subchapter, we will explore two key tactics to achieve this: paid advertising and cross-promotion and partnerships.

Paid Advertising

Paid advertising is an effective way to reach a wide audience, drive app installs, and increase visibility in a short period. By investing in paid advertising, you can target specific user segments, track your campaign's performance, and optimize your marketing budget. Here are some key considerations when implementing paid advertising for your app:

1.   Choose the right platforms: Several platforms offer app advertising, including social media networks like Facebook, Instagram, and Twitter, as well as ad networks like Google Ads and Apple Search Ads. Research and choose the platforms that best suit your target audience and budget.
2.   Define your target audience: To maximize the effectiveness of your paid advertising campaigns, you must clearly define your target audience. Consider factors such as demographics, interests, and behaviors to create user personas that will guide your ad targeting.
3.   Set a budget and goals: Before launching your paid advertising campaigns, set a budget and determine your goals, such as the number of app installs or the cost per install (CPI). Having a clear budget and objectives will help you measure the success of your campaigns and optimize your ad spend.

4. Create compelling ad creatives: Design eye-catching ad creatives that showcase your app's unique features and benefits. Use a mix of images, videos, and text to appeal to different user preferences and increase the likelihood of users engaging with your ads.

5. Test and optimize: Continuously test different ad creatives, targeting options, and bidding strategies to find the most effective combination for your app. Regularly monitor your campaigns' performance and make data-driven adjustments to maximize your return on investment (ROI).

Cross-Promotion and Partnerships

Another effective user acquisition strategy is cross-promotion and partnerships. This tactic involves collaborating with other app developers, businesses, or influencers to promote your app to their user base, and vice versa. By leveraging each other's audiences, both parties can benefit from increased visibility and user acquisition.

1. Identify potential partners: Look for businesses, apps, or influencers that have a complementary target audience and share similar values or goals. Research their user base, reach, and engagement to ensure they align with your app's objectives.

2. Approach potential partners: Reach out to potential partners with a tailored pitch that highlights the mutual benefits of a partnership. Be clear about your expectations and what you can offer in return, such as promotional support, revenue sharing, or access to exclusive content or features.

3. Develop a cross-promotion strategy: Work with your partner to develop a cross-promotion strategy that benefits both parties. This could include in-app promotions, co-branded marketing campaigns, or social media shoutouts. Ensure that the cross-promotion is relevant and adds value to both parties' user bases.

4. Measure and analyze results: Track the performance of your cross-promotion efforts using relevant metrics, such as user acquisition, engagement, and revenue. Use this data to optimize your partnership strategy and identify new opportunities for collaboration.

5.   Maintain and expand partnerships: Nurture your existing partnerships and look for new opportunities to expand your network. By building long-term relationships with partners, you can create a strong support system that can help drive ongoing user acquisition and growth for your app.

In conclusion, implementing a robust user acquisition strategy is essential for driving app downloads, engagement, and revenue. By leveraging paid advertising and cross-promotion and partnerships, you can reach a wider audience and attract new users to your app. As you execute your user acquisition strategy, continuously monitor, test, and optimize your efforts to maximize your ROI and ensure long-term success for your app business.

The key takeaways for user acquisition through paid advertising and cross-promotion and partnerships include:

- Choosing the right platforms for your paid advertising campaigns and creating compelling ad creatives to attract users.
- Defining your target audience, setting a budget, and outlining clear goals to guide your paid advertising efforts.
- Identifying and approaching potential partners that share a complementary target audience and align with your app's objectives.
- Developing and executing a cross-promotion strategy that benefits both parties and adds value to their user bases.
- Continuously measuring and analyzing the results of your user acquisition efforts to optimize your strategies and identify new opportunities for growth.

By combining these two powerful user acquisition tactics, you can effectively drive app downloads and build a loyal user base that will contribute to your app's overall success. As you continue to grow your app business, remain agile and responsive to market trends and user feedback, and be prepared to adapt your user acquisition strategies as needed. With a strong foundation in place and a commitment to ongoing optimization and growth, you can turn your app idea into a profitable business and thrive in the competitive app marketplace.

CHAPTER FOUR

In-App Advertising: Maximizing Revenue Without Sacrificing User Experience

Understanding Ad Formats

In this part, we will explore various ad formats available for app developers, focusing on three popular formats: banner ads, interstitial ads, and rewarded ads. By understanding these formats and their characteristics, you can make informed decisions about which type of ads best align with your app's user experience and monetization goals.

Banner Ads

Banner ads are the most common form of in-app advertising, known for their simplicity and ubiquity across mobile platforms. These ads are usually rectangular in shape and placed at the top or bottom of the app's interface. Banner ads can be static or animated and typically promote other apps or products relevant to your app's target audience.

Advantages of banner ads include their non-intrusive nature, as they generally do not disrupt the user experience. They can also be easily implemented within the app's layout and can generate revenue on a cost-per-impression (CPM) or cost-per-click (CPC) basis.

However, banner ads also have their drawbacks. Due to their smaller size and placement, they can sometimes go unnoticed by users, leading to lower click-through rates (CTR) and, consequently, reduced revenue. Additionally, the prevalence of banner ads has led to banner blindness, where users subconsciously ignore the ads, further reducing their effectiveness.

Interstitial Ads

Interstitial ads are full-screen advertisements that appear during natural breaks in the app's user experience, such as between levels in a game or during a pause in content consumption. They can be static, video, or interactive and are designed to capture users' attention more effectively than banner ads.

The primary advantage of interstitial ads is their ability to generate higher CTRs and eCPMs (effective cost per thousand impressions) due to their prominent placement and larger size. They can also be a more effective way to promote app installs, as users are more likely to engage with a full-screen ad.

However, interstitial ads can also negatively impact the user experience if not implemented properly. If they appear too frequently or at inopportune moments, users may become frustrated and abandon your app. It's crucial to strike the right balance between monetization and user experience by carefully considering the timing and frequency of interstitial ads.

Rewarded Ads

Rewarded ads, often referred to as incentivized ads, are a popular ad format that offers users a reward, such as in-app currency, premium content, or additional app features, in exchange for watching a video ad or completing an in-app action (e.g., signing up for a newsletter). Rewarded ads are most commonly used in mobile games but can also be applied in other app genres.

The primary advantage of rewarded ads is their ability to enhance the user experience while generating revenue. By offering a tangible benefit to users, they are more likely to engage with the ad, leading to higher eCPMs and retention rates. Rewarded ads also benefit from a positive perception, as users view them as a means to access premium content without spending real money.

However, rewarded ads also have their drawbacks. Over-reliance on this ad format can lead to users becoming dependent on rewards, potentially reducing the effectiveness of other monetization strategies such as in-app purchases. Additionally, it's crucial to ensure that the rewards offered are valuable to users and do not unbalance the app's overall economy or experience.

In conclusion, selecting the appropriate ad format for your app is crucial in maximizing revenue without sacrificing user experience. Each format has its own advantages and drawbacks, and understanding their characteristics will help you make informed decisions about the best fit for your app. By strategically implementing and optimizing these ad formats, you can create a sustainable revenue stream while maintaining a high-quality user experience, setting the stage for long-term success in the competitive app market.

Ad Network Selection and Integration

Evaluating Ad Networks

The selection of the right ad network is crucial to the success of your in-app advertising strategy. With numerous ad networks available, it's essential to evaluate them based on specific criteria to ensure they align with your app's needs and monetization goals. In this sub-subchapter, we will discuss the key factors to consider when evaluating ad networks, including their reputation, fill rate, eCPM, payment terms, and targeting capabilities.

1. Reputation: A reputable ad network will have a proven track record of success in the industry, with satisfied clients and a history of timely payments. Research each ad network you're considering, read reviews and testimonials from other app developers, and seek recommendations from your professional network. Partnering with a reputable ad network can significantly impact your revenue generation and overall user experience.

2. Fill Rate: The fill rate is the percentage of ad requests that the ad network successfully fills with an ad. A high fill rate is

desirable, as it ensures your ad inventory will be effectively utilized, maximizing your revenue potential. However, keep in mind that the fill rate can vary based on factors such as the user's location, device, and connection type. When evaluating ad networks, inquire about their historical fill rates and compare them to industry benchmarks.

3. eCPM (Effective Cost per Mille): eCPM is a critical metric that reflects the average revenue generated per 1000 ad impressions. A higher eCPM indicates that the ad network's ads are more effective at driving revenue. When comparing ad networks, consider their average eCPM rates to help you determine which one will likely generate the most revenue for your app.

4. Payment Terms: It's essential to understand the payment terms offered by each ad network, including the payment frequency, payment methods, and minimum payment thresholds. Choose an ad network with payment terms that align with your cash flow needs and preferences.

5. Targeting Capabilities: Effective ad targeting is crucial to delivering relevant ads that will engage your users and drive higher CTRs and conversion rates. Look for ad networks that offer advanced targeting options, such as demographics, interests, behaviors, and device types. These capabilities will enable you to serve ads that resonate with your target audience, ultimately improving the overall user experience.

Implementing Ads in Your App

Once you've selected an ad network that meets your criteria, the next step is to integrate the ads into your app. In this sub-subchapter, we will discuss best practices for implementing ads to maximize revenue without sacrificing user experience, including proper ad placement, frequency, and testing.

1. Ad Placement: The placement of ads within your app will significantly impact their effectiveness and the user experience. Choose placements that are visible but not intrusive, ensuring that users can still engage with your app's content and features without being overwhelmed by ads. Consider using native ads or other ad formats that blend

seamlessly with your app's interface and design.

2.   Ad Frequency: Balancing the frequency of ads is crucial to maintaining a positive user experience while maximizing revenue. Showing too many ads can lead to user frustration and abandonment, while too few ads can result in missed revenue opportunities. Use analytics to monitor user engagement and retention rates, adjusting the frequency of ads as needed to optimize the balance between revenue generation and user satisfaction.

3.   A/B Testing: Implementing A/B testing can help you determine the most effective ad formats, placements, and frequencies for your app. By testing different ad configurations, you can identify the optimal setup that drives the highest revenue without negatively impacting the user experience. Continuously refine your ad strategy based on testing data and user feedback to ensure long-term success.

4.   Ad Mediation: Using an ad mediation platform can help you maximize revenue by managing multiple ad networks and optimizing ad delivery. Ad mediation platforms automatically select the highest paying ads from your integrated ad networks, ensuring that you're always generating the most revenue possible from your ad inventory. They also simplify the process of integrating and managing multiple ad networks, saving you time and effort.

5.   Monitoring Performance: Regularly monitoring the performance of your ads is critical to optimizing your monetization strategy. Keep an eye on metrics such as eCPM, fill rate, click-through rate (CTR), and conversion rate to gauge the effectiveness of your ads. Analyzing this data will help you identify areas for improvement and make informed decisions about ad placements, formats, and targeting.

6.   Maintaining User Experience: As you implement ads in your app, it's essential to prioritize user experience to ensure long-term success. Regularly gather user feedback through surveys, app store reviews, and in-app prompts to understand how users feel about the ads they're encountering. Use this feedback to make adjustments to your ad strategy, ensuring that ads are not hindering users' enjoyment of your app.

7.   Compliance with Ad Policies: Adherence to the policies and guidelines set forth by your chosen ad networks and app

stores is essential to avoid any potential issues or penalties. Familiarize yourself with these policies, and ensure your app and ads are in compliance. This will not only protect your app from potential removal from app stores but will also help maintain a high-quality user experience.

In conclusion, implementing ads in your app is a crucial aspect of app monetization. By carefully evaluating and selecting the right ad networks, and effectively integrating ads into your app while maintaining a focus on user experience, you can maximize revenue generation and set your app on the path to long-term success. Keep iterating and optimizing your strategy based on data and user feedback to ensure continued growth and profitability.

Ad Placement and Optimization

Strategic Ad Placement

When integrating ads into your app, one of the most critical aspects to consider is the placement of these ads. Strategic ad placement ensures that your ads are visible and engaging, without disrupting the user experience. It is crucial to strike a balance between maximizing revenue and maintaining a positive user experience to encourage user retention and long-term success.

To effectively place ads in your app, consider the following guidelines:

1. Understand user behavior: Analyze user engagement patterns within your app to identify the most appropriate locations for ad placement. Place ads in natural breaks within the app's content or during moments when users are likely to be receptive to ads, such as between levels in a game or after completing a task.
2. Avoid obtrusive placements: Ads should never obstruct the user's primary interaction with the app. Avoid placing ads in areas where users may accidentally click on them while trying

to interact with your app's core functionality.

3. Leverage native ads: Native ads seamlessly integrate with your app's design and content, creating a more cohesive user experience. By blending with the surrounding content, native ads are less disruptive and can lead to higher engagement rates.

4. Test different ad formats: Experiment with various ad formats (such as banner, interstitial, and rewarded ads) to determine which ones work best for your app and audience. Different formats may be more suitable for different sections of your app or user segments.

A/B Testing and Iteration

Once you have strategically placed ads within your app, it's essential to continuously test and optimize their performance. A/B testing is a valuable method to compare different ad placements, formats, and designs to determine which performs best for your specific app and audience.

Here are some essential steps for conducting A/B testing and iterative optimization in your app:

1. Identify your testing variables: Determine the elements you want to test, such as ad placement, ad format, or ad design. For example, you may want to compare the performance of banner ads at the top versus the bottom of the screen or test different designs for native ads.

2. Develop a hypothesis: Based on your understanding of user behavior and preferences, develop a hypothesis about which ad variations will perform better. This hypothesis will serve as a guide for your testing process and help you interpret the results.

3. Split your audience: Divide your app's user base into two or more groups, ensuring that each group is representative of your overall audience. Assign each group a different ad variation to test.

4. Monitor performance metrics: Collect data on key performance metrics, such as click-through rate (CTR), conversion rate, and eCPM (effective cost per mille). This data will help you

determine the effectiveness of each ad variation and identify the best-performing option.

5. Analyze the results: Once you have collected enough data, analyze the results to determine if there is a statistically significant difference between the performance of the ad variations. If one variation outperforms the others, you may choose to implement it across your entire app. If the results are inconclusive or the differences are minimal, consider running additional tests with new variables.

6. Iterate and optimize: Continue to test and optimize your ad placements, formats, and designs to maximize revenue and maintain a positive user experience. As your app and audience evolve, regularly reassess your ad strategy to ensure that it remains effective and relevant.

In summary, strategic ad placement and continuous A/B testing and iteration are essential components of a successful in-app advertising strategy. By carefully considering the placement of ads within your app and consistently testing and optimizing their performance, you can maximize revenue generation while maintaining a positive user experience. As you refine your monetization strategy, always prioritize user feedback and data-driven insights to ensure long-term success for your app and business.

CHAPTER FIVE

In-App Purchases: Encouraging User Spending

Crafting Compelling In-App Offers

Virtual Goods and Currency

In-app purchases (IAP) can be a highly effective monetization strategy when executed correctly. One of the most popular forms of IAP is the sale of virtual goods and currency, which can enhance the user experience and encourage in-app spending. Virtual goods can include items, power-ups, or other enhancements that users can buy to improve their in-app experience or performance. Virtual currency, on the other hand, serves as a medium of exchange within the app and can be used to purchase virtual goods or other in-app content.

To create compelling virtual goods and currency offers, consider the following strategies:

1. Design desirable virtual goods: Ensure that your virtual goods are visually appealing and hold value for your users. They should enhance the user experience, providing a sense of accomplishment or customization. For example, in a mobile game, you might offer power-ups that help users progress through levels more quickly or exclusive skins for their characters.

2. Balance between attainability and exclusivity: While some virtual goods should be easily attainable through in-app

currency or gameplay, others should be more exclusive to incentivize spending. Exclusive items can be offered for a limited time, in limited quantities, or require a significant amount of virtual currency to purchase, driving users to spend real money on currency packs.

3. Employ psychological pricing techniques: Utilize pricing strategies that encourage users to spend more, such as offering virtual currency packs with bonus amounts for higher spending or using tiered pricing structures for virtual goods.

4. Implement bundles and sales: Offer bundles of virtual goods or currency at discounted rates to encourage users to make larger purchases. Limited-time sales can create a sense of urgency, encouraging users to spend before the offer expires.

Premium Features and Content

Another effective IAP strategy is to offer premium features and content to users who are willing to pay for an enhanced experience. This can include unlocking additional app functionality, providing access to exclusive content, or removing ads. To craft compelling premium features and content offers, consider the following approaches:

1. Identify valuable features and content: Analyze your users' needs and preferences to determine which features or content are most valuable to them. This might include advanced app functionality, additional content such as levels or challenges, or ad-free experiences.

2. Offer a clear value proposition: Ensure that your premium features and content offer a clear and compelling value proposition to your users. The benefits should be easily understood and provide a strong incentive for users to upgrade. For example, a music streaming app might offer an ad-free experience and offline listening capabilities for premium subscribers.

3. Implement tiered pricing: Offer different pricing tiers for premium features and content, catering to various user segments and their willingness to pay. This can help maximize revenue by appealing to users with different budgets and preferences.

4. Leverage free trials and limited access: Offer users the opportunity to try premium features or content for a limited time, allowing them to experience the benefits firsthand before committing to a purchase. This can help users understand the value of the premium offering and increase the likelihood of converting to a paid user.

5. Use targeted marketing and promotions: Promote your premium features and content to users who are most likely to be interested in them. Use data-driven insights to identify and target users with personalized promotions, highlighting the specific features or content that are most relevant to their interests and needs.

By carefully crafting compelling virtual goods, currency offers, and premium features and content, you can effectively drive in-app spending and maximize revenue generation. It's essential to continuously test and optimize your offers, utilizing data-driven insights to refine your approach and better cater to your users' needs and preferences. When executed correctly, in-app purchases can serve as a powerful monetization strategy that not only generates revenue but also enhances the user experience and fosters long-term user engagement.

Pricing Strategies for In-App Purchases

Tiered Pricing

Tiered pricing is an effective strategy for in-app purchases (IAP) that allows you to cater to different user segments and their willingness to pay. By offering multiple pricing tiers for your virtual goods, currency, or premium features, you can increase conversion rates, maximize revenue, and improve the overall user experience. The key to successfully implementing tiered pricing lies in understanding your target audience, crafting compelling offers, and continuously optimizing your pricing structure based on data-driven insights.

Consider the following steps for implementing a successful tiered pricing strategy:

* * *

1. Analyze user behavior and preferences: To create an effective tiered pricing structure, begin by analyzing your users' behavior, preferences, and willingness to pay. This can be achieved through in-app analytics, user feedback, and market research. Identifying common user segments and their needs will help you craft pricing tiers that cater to different budgets and preferences.

2. Create multiple pricing tiers: Develop multiple pricing tiers, each offering varying levels of value and benefits. This may include different quantities of virtual goods or currency, access to varying levels of premium content, or varying degrees of app functionality. Ensure that the tiers are well-differentiated and cater to different user segments.

3. Offer a clear value proposition: Each pricing tier should offer a clear and compelling value proposition to users. Make sure the benefits of upgrading to a higher tier are easily understood and provide a strong incentive for users to make a purchase. For example, a tiered subscription plan for a video streaming app might offer different levels of content access, streaming quality, and concurrent device usage.

4. Utilize psychological pricing techniques: Incorporate pricing techniques that encourage users to spend more, such as anchoring or decoy pricing. Anchoring involves setting a higher-priced option as a reference point, making lower-priced options appear more attractive. Decoy pricing involves introducing a less attractive option to make other options seem more valuable by comparison.

5. Continuously test and optimize: Regularly test and optimize your tiered pricing structure based on user feedback, conversion rates, and revenue generation. This may involve adjusting pricing, benefits, or the number of tiers offered. Use data-driven insights to refine your approach and better cater to your users' needs and preferences.

Limited-Time Offers

Limited-time offers (LTOs) are another powerful pricing strategy for in-app purchases that can help drive user spending, create a sense of urgency, and boost overall revenue. By offering discounts, exclusive content, or bundled items for a limited time, you can encourage users to make impulsive purchases, increasing conversion rates and enhancing the user experience.

Consider the following steps for implementing a successful limited-time offer strategy:

1. Identify compelling offers: Begin by identifying the types of offers that will resonate with your target audience. This might include discounted virtual goods or currency, exclusive content or features, or limited-time bundles. Consider your users' preferences and the unique value proposition of your app when crafting your offers.
2. Create a sense of urgency: To drive user spending, it's essential to create a sense of urgency around your limited-time offers. This can be achieved through countdown timers, limited quantities, or exclusive access for a specified period. Make sure users are aware of the time-sensitive nature of the offer and the benefits they stand to gain by acting quickly.
3. Promote your limited-time offers: Effectively promoting your LTOs is crucial to their success. Use in-app notifications, push notifications, and email campaigns to inform users of the offer and encourage them to take advantage. Leverage social media and other marketing channels to amplify your message and reach a broader audience.
4. Measure and analyze results: Track the performance of your limited-time offers by monitoring key metrics such as conversion rates, revenue generated, and user engagement. Use this data to identify which offers resonate most with your target audience and make data-driven decisions for future campaigns.
5. Learn from past offers: After each limited-time offer campaign, analyze the results to identify trends, successes, and areas for improvement. This information can help you refine your LTO strategy, making future offers more effective and better aligned with your users' needs and preferences.
6. Test different types of offers: Experiment with various types of limited-time offers to determine what works best for your app and audience. This might include testing different discounts, exclusive content, or bundle configurations. Continuously

iterate and optimize your offers based on data-driven insights and user feedback.

7. Leverage holidays and special events: Capitalize on holidays, seasonal events, or other relevant occasions to create themed limited-time offers. These events can provide a natural opportunity for users to engage with your app and make purchases. For example, offer discounted in-app currency during the holiday season or release exclusive content to coincide with a major industry event.

8. Balance limited-time offers with long-term value: While LTOs can be an effective way to drive short-term revenue, it's essential to balance these promotions with strategies that foster long-term user engagement and value. Avoid over-relying on limited-time offers, as this can lead to user fatigue and diminished long-term engagement. Instead, use LTOs in conjunction with other monetization strategies, such as tiered pricing, in-app advertising, or subscriptions, to create a well-rounded approach that maximizes revenue and user satisfaction.

By implementing tiered pricing and limited-time offers, you can craft a comprehensive pricing strategy that caters to different user segments, drives in-app spending, and ultimately, increases revenue generation. The key to success lies in understanding your target audience, creating compelling offers, and continuously testing and optimizing your approach based on data-driven insights. When executed effectively, these pricing strategies can help you maximize the revenue potential of your app and create a profitable business while maintaining a positive user experience.

Enhancing User Retention and Engagement

Loyalty Programs and Rewards

Creating a loyal user base is essential for the long-term success of any app. One way to encourage user retention and increase in-app spending is by implementing loyalty programs and rewards systems. These programs incentivize users to keep using your app and make purchases, leading to increased revenue and a more engaged user base.

1. Types of loyalty programs: There are several types of loyalty programs that can be implemented within an app. Some common examples include point-based systems, where users earn points for completing certain actions, such as making a purchase or reaching a specific milestone. These points can then be redeemed for rewards, such as in-app currency, premium content, or discounts. Other loyalty programs include tiered systems, where users are rewarded with exclusive benefits based on their level of engagement or spending within the app.

2. Personalized rewards: To make your loyalty program more effective, tailor rewards to the preferences and interests of your users. By offering personalized rewards, you increase the perceived value of the program and encourage users to engage more with your app. Collect data on user behavior, such as purchase history and in-app activity, to inform the design of personalized rewards and ensure they resonate with your target audience.

3. Gamification: Integrating gamification elements into your loyalty program can further enhance user engagement and retention. Gamification involves incorporating game-like elements, such as challenges, leaderboards, and badges, into non-gaming contexts. By making the process of earning rewards more fun and interactive, you can motivate users to continue using your app and make in-app purchases.

4. Promoting your loyalty program: Ensure that users are aware of your loyalty program and its benefits. Use in-app messaging, push notifications, and other marketing channels to promote the program and encourage users to participate. Highlight the value of the rewards and emphasize how easy it is to earn points or advance in the program.

Push Notifications and In-App Messaging

Push notifications and in-app messaging are powerful tools for encouraging user engagement and retention. By delivering targeted and relevant messages to users at the right time, you can increase the likelihood that they will return to your app and make in-app purchases.

1.   Push notifications: Push notifications are messages that appear on a user's device, even when the app is not in use. They can be used to alert users to new content, promotions, or other app-related news. When used effectively, push notifications can drive user engagement, increase app usage, and boost in-app spending.

2.   In-app messaging: In-app messaging refers to messages displayed within the app while the user is actively using it. These messages can be used to promote special offers, suggest personalized recommendations, or provide helpful tips and tutorials. In-app messaging can be particularly effective for driving user engagement and in-app purchases, as users are already in the context of using the app when they receive the message.

3.   Personalization and relevance: To maximize the effectiveness of your push notifications and in-app messaging, ensure that the content is personalized and relevant to each user. Use data on user behavior, interests, and preferences to create tailored messages that resonate with your target audience. This will increase the likelihood that users will engage with your messages and take the desired action, such as making a purchase or participating in a loyalty program.

4.   Timing and frequency: The timing and frequency of your push notifications and in-app messages can significantly impact their effectiveness. Too many messages can lead to notification fatigue, causing users to ignore or even disable notifications from your app. On the other hand, too few messages can result in missed opportunities to engage users and drive in-app spending. Test different timings and frequencies to find the optimal balance for your app and audience.

5.   A/B testing: Continuously test and iterate on your push notification and in-app messaging strategy to ensure optimal performance. A/B testing involves creating multiple variations of a message, such as different headlines, images, or call-to-action buttons, and then measuring the engagement and conversion rates for each variation. By analyzing the performance of each variation, you can identify the most effective messaging strategies and make data-driven decisions

to improve your overall approach.

6. Segmentation: Segmenting your user base allows you to deliver more targeted and relevant messages to specific groups of users. This can result in higher engagement and conversion rates compared to a one-size-fits-all messaging approach. Consider segmenting users based on factors such as demographics, in-app behavior, purchase history, or engagement level. For example, you might send a special promotion to users who have not made a purchase in the past 30 days or a helpful tutorial to new users who have not yet completed the onboarding process.

7. Analytics and measurement: Establish key performance indicators (KPIs) to measure the success of your push notification and in-app messaging strategy. Common KPIs include open rates, click-through rates, conversion rates, and revenue generated from messages. By tracking these metrics, you can identify areas for improvement and make data-driven decisions to optimize your messaging strategy over time.

8. Compliance with regulations: When using push notifications and in-app messaging, it's essential to comply with relevant regulations and guidelines, such as the General Data Protection Regulation (GDPR) and the California Consumer Privacy Act (CCPA). Ensure that users have provided the necessary consent to receive messages, and provide options for users to manage their notification preferences or opt-out of receiving messages if desired.

In conclusion, loyalty programs and rewards, as well as push notifications and in-app messaging, are powerful strategies for enhancing user retention and engagement in your app. By implementing these strategies, you can drive in-app spending and maximize your app's revenue potential. Remember to continually test, iterate, and optimize your approach to ensure the best possible results for your app and business.

CHAPTER SIX

Subscription Models: Creating Recurring Revenue

Determining Subscription Tiers and Benefits

As the app industry continues to evolve, subscription models have emerged as a popular and effective way for developers to generate recurring revenue. By offering users ongoing access to exclusive content and features in exchange for a regular fee, you can create a sustainable and scalable business model. In this subchapter, we will explore how to determine subscription tiers and benefits, focusing on the Basic, Pro, and Premium plans, and the exclusive content and features that can be offered at each level.

Basic, Pro, and Premium Plans

One of the first steps in developing a subscription model is to determine the structure of your subscription tiers. A common approach is to offer three tiers: Basic, Pro, and Premium. This three-tiered structure provides users with options, allowing them to choose the plan that best fits their needs and budget. Each tier should offer a distinct set of benefits, with higher tiers providing more value and exclusive content or features.

Basic Plan: The Basic plan is typically the lowest-priced option and is aimed at users who want access to your app's core functionality but may not require advanced features or additional content. This plan should offer a solid user experience and sufficient value to justify the subscription fee while also serving as an entry point for users who might eventually upgrade to higher tiers. The Basic plan can include access to the primary content library, basic features, and limited customer support.

Pro Plan: The Pro plan is a mid-tier option, providing additional benefits and exclusive features to users who are willing to pay a higher fee. This plan should offer a more comprehensive experience than the Basic plan, including advanced features, additional content, or enhanced customer support. The Pro plan can also include access to a larger content library, the ability to customize the app experience, or even a reduced ad experience.

Premium Plan: The Premium plan is the highest-priced option and is targeted at users who are seeking the most complete and exclusive experience. This plan should offer a significant value proposition, providing users with access to all content and features, including those not available in the lower tiers. The Premium plan can include benefits such as an ad-free experience, priority customer support, early access to new content or features, and exclusive bonuses or rewards.

Exclusive Content and Features

To justify the price difference between subscription tiers and encourage users to upgrade, each tier should offer a unique set of exclusive content and features. This can include access to additional content, advanced functionality, enhanced customization options, and other benefits that provide value to subscribers.

Content Libraries: Offer a tiered content library, with more extensive or exclusive content available at higher subscription levels. For example, a streaming app might provide access to a basic library of films and TV shows for Basic subscribers, while Pro subscribers gain access to a larger library with additional content, and Premium subscribers receive exclusive access to original programming or early releases.

Advanced Features: Provide advanced features that cater to the specific needs of your target audience. For instance, a photo editing app might offer basic editing tools to Basic subscribers, while Pro subscribers gain access to more advanced features like professional filters, layers, or image manipulation tools, and Premium subscribers receive exclusive access to tools like AI-powered image enhancement or batch processing.

Customization Options: Enhance the user experience by offering customization options at higher subscription levels. For example, a productivity app could provide Basic subscribers with standard templates and themes, while Pro subscribers gain access to additional themes and the ability to customize their workspace, and Premium subscribers receive exclusive access to custom templates and integration options.

Ad-Free Experience: Offer an ad-free experience as a benefit for higher-tier subscribers, removing advertisements and providing a more seamless user experience. For example, a music streaming app might offer Basic subscribers access to the content library with occasional advertisements, while Pro subscribers have fewer ads and Premium subscribers enjoy an entirely ad-free experience.

Priority Customer Support: Provide enhanced customer support options for higher-tier subscribers. This can include offering faster response times, dedicated support channels, or even one-on-one support sessions. By delivering premium support to your most valuable subscribers, you can foster loyalty and improve user satisfaction.

Early Access to New Content or Features: Reward higher-tier subscribers with early access to new content or features before they become available to other users. This exclusive access can create a sense of prestige and exclusivity, encouraging users to maintain their subscription and explore the new content or features before anyone else.

Exclusive Bonuses or Rewards: Offer exclusive bonuses or rewards for higher-tier subscribers, such as in-app currency, discounts on future purchases, or access to members-only events. These perks can enhance the perceived value of the subscription and motivate users to maintain their subscription status.

In conclusion, creating a successful subscription model involves determining the right structure for your subscription tiers and offering exclusive content and features that appeal to your target audience. By providing users with a range of options and tailoring your offerings to their needs and budgets, you can maximize revenue and encourage long-term user engagement.

Keep in mind that it's essential to continuously analyze user behavior and feedback to optimize your subscription model and ensure it remains relevant and valuable to your audience. With a well-crafted subscription model, you can create a sustainable and profitable app business that delivers ongoing value to your users.

Pricing and Billing Strategies

As a software developer and entrepreneur with for 15+ years, I have witnessed the power of subscription models in creating recurring revenue for app businesses. In this chapter, we will delve into the crucial aspect of pricing and billing strategies, focusing on monthly vs. annual billing, as well as free trials and discounts. These elements can significantly impact your app's revenue, user growth, and long-term success.

Monthly vs. Annual Billing

Determining the most appropriate billing cycle for your app's subscription model is an essential step in maximizing revenue and user satisfaction. Both monthly and annual billing options have their advantages and drawbacks.

Monthly Billing:

Advantages: Monthly billing is generally more attractive to users due to its lower upfront cost, allowing them to try your app without a significant financial commitment. This can result in a higher conversion rate from free users to paying subscribers. Additionally, monthly billing provides users with the flexibility to cancel their subscription at any time, which can be perceived as more consumer-friendly.

Drawbacks: Monthly billing can lead to higher churn rates as users may cancel their subscription more frequently. This may require you to invest more in user acquisition and retention efforts. Furthermore, the administrative and payment processing costs associated with monthly billing can be higher.

Annual Billing:

Advantages: Annual billing offers users a discounted rate in exchange for a longer commitment, which can result in higher customer lifetime value (CLTV) and more predictable revenue streams. This option also reduces payment processing and administrative costs due to the less frequent billing cycle.

Drawbacks: The higher upfront cost of annual billing can be a barrier to entry for some users, potentially leading to a lower conversion rate. Moreover, annual billing may not be suitable for apps with seasonal usage patterns, as users may not see the value in paying for a full year.

In general, offering both monthly and annual billing options can cater to a wider range of user preferences, maximizing revenue and user satisfaction. You may also consider other billing frequencies, such as quarterly or bi-annual plans, depending on your app's specific use case and target audience.

Free Trials and Discounts

Offering free trials and discounts can be an effective strategy for attracting new users, converting them into paying subscribers, and encouraging existing users to upgrade their subscription plans.

Free Trials:

Advantages: Free trials allow users to experience the premium features of your app without any financial commitment. This can help to showcase the value of your app and convert users into paying subscribers. A well-executed free trial can lead to increased user acquisition, higher conversion rates, and ultimately, more revenue.

Drawbacks: Free trials can be subject to abuse by users who repeatedly sign up with different accounts to access premium features for free. Additionally, free trials may not be suitable for all apps, especially those with a high cost of providing the service or those that offer a one-time benefit.

When offering a free trial, consider the optimal duration (e.g., 7 days, 14 days, or 30 days) and ensure that users have a clear understanding of the trial's terms and conditions. It's crucial to communicate the value of your app's premium features during the trial period and provide a seamless conversion process for users to become paying subscribers.

Discounts:

Advantages: Discounts can incentivize users to upgrade their subscription plan, commit to a longer billing cycle, or renew their subscription. Offering discounts can also be an effective strategy for seasonal promotions, targeting specific user segments, or as a reward for user loyalty.

Drawbacks: Overuse of discounts can lead to a perceived reduction in the value of your app or train users to wait for discounts before making a purchase. Moreover, discounts can temporarily reduce your app's revenue per user, which may impact your overall profitability if not carefully managed.

When implementing discounts, consider the following best practices:

Timing: Choose the right time for offering discounts, such as during holiday seasons, special events, or when your app experiences a significant increase in user traffic. Be cautious not to offer discounts too frequently, as this may lead to users expecting them and waiting for the next promotion.

* * *

Targeting: Segment your user base and offer targeted discounts to specific groups, such as new users, long-time users, or users who have been inactive for a certain period. This can help you maximize the impact of your discounts while minimizing the risk of revenue loss.

Exclusivity: Create a sense of exclusivity by offering discounts to a limited number of users, for a limited time, or as part of a special promotion. This can help drive urgency and encourage users to take advantage of the offer before it expires.

Value: Ensure that the discount provides genuine value to the user, and it's not just a superficial price reduction. Make sure that users understand the benefits they will receive when they subscribe or upgrade, and highlight the savings they will make compared to the regular price.

In conclusion, determining the optimal pricing and billing strategies for your app's subscription model is crucial for maximizing revenue and user satisfaction. By carefully considering the advantages and drawbacks of different billing cycles, free trials, and discounts, you can create a compelling and effective monetization strategy that encourages users to spend within your app and remain loyal customers.

As an experienced software developer and entrepreneur, I understand the importance of striking the right balance between revenue generation and providing an excellent user experience. By implementing the strategies outlined in this chapter, you can set your app on the path to monetization success while keeping your users engaged and satisfied.

Retaining and Upselling Subscribers

Dunning Management and Failed Payments
In the world of subscription-based app monetization, managing

failed payments and dunning processes is essential to maintaining a healthy revenue stream. Dunning refers to the practice of communicating with subscribers about their failed payments and taking steps to resolve the issue. As a seasoned software developer and entrepreneur, I understand the importance of establishing a robust dunning management system to minimize the impact of failed payments on your app's revenue.

Failed payments can occur for various reasons, including expired or canceled credit cards, insufficient funds, or technical issues with the payment gateway. To address these issues effectively, implement the following best practices:

1. Set up automated notifications: Configure your billing system to send automated email notifications to subscribers as soon as a payment fails. These notifications should inform the subscriber about the issue, provide guidance on resolving it, and include a clear call-to-action to update their payment information.

2. Customize the messaging: Personalize your dunning emails with the subscriber's name and specific information about the failed payment. Ensure that your message is concise, polite, and professional, and avoid using aggressive or threatening language.

3. Offer multiple payment options: Provide subscribers with alternative payment methods to minimize the risk of failed payments. For example, consider offering PayPal or other popular payment gateways alongside credit card payments.

4. Implement a retry schedule: Configure your billing system to automatically retry failed payments at predetermined intervals, such as every few days or weekly. This helps increase the chances of capturing the payment without requiring any additional action from the subscriber.

5. Monitor and analyze failed payments: Regularly review your failed payment data to identify trends and patterns. This can help you identify any recurring issues, such as specific payment gateways causing a high rate of failures or specific subscription plans experiencing more frequent issues.

Subscriber Engagement and Value

Retaining subscribers and encouraging them to upgrade to higher-tier plans is critical to maximizing your app's revenue potential. The key to achieving this lies in maintaining high levels of subscriber engagement and demonstrating the value of your subscription offerings. Consider the following strategies to achieve this:

1. Offer exclusive content and features: Provide subscribers with access to unique features, content, or services that are not available to free users. This helps reinforce the value of their subscription and incentivize them to remain subscribed or upgrade to higher-tier plans.

2. Maintain a regular content release schedule: Ensure that your app is constantly updated with fresh content or features, giving subscribers a reason to stay engaged and continue using your app. A consistent content release schedule also helps create a sense of anticipation and excitement among your subscribers.

3. Communicate with subscribers regularly: Send periodic email newsletters or in-app notifications to keep subscribers informed about new content, features, or upcoming events. This helps maintain engagement and reminds users of the ongoing value they receive from their subscription.

4. Offer personalized recommendations: Use data-driven insights to recommend content, features, or services that are tailored to each subscriber's preferences and usage patterns. This not only enhances the user experience but also reinforces the value of their subscription.

5. Leverage gamification and social features: Encourage users to interact with each other within your app and incorporate gamification elements such as leaderboards, achievements, or rewards. This can foster a sense of community and competition, further driving engagement and retention.

6. Monitor and analyze user behavior: Keep a close eye on your app's analytics to identify trends and patterns in subscriber behavior. Use these insights to make data-driven decisions on improving your app's content, features, or user experience.

By implementing these strategies, you can effectively retain and upsell subscribers, maximizing your app's revenue potential. Balancing dunning management with subscriber engagement and value is essential to creating a successful subscription model. As a software developer and entrepreneur with two decades of experience, I know that these tactics, when implemented correctly, can significantly impact your app's profitability and long-term sustainability.

1. Offer exceptional customer support: Providing top-notch customer support is crucial for maintaining subscriber satisfaction and loyalty. Ensure that your support team is responsive, knowledgeable, and proactive in addressing customer concerns. This not only helps in resolving issues promptly but also enhances your app's reputation and demonstrates your commitment to providing value to your subscribers.

2. Implement a referral program: Encourage your existing subscribers to recommend your app to their friends, family, or colleagues by offering incentives such as discounts or bonus content. This not only helps to attract new subscribers but also increases the perceived value of your app to your current subscribers.

3. Test different upselling techniques: Experiment with various upselling tactics, such as offering discounted upgrades or time-sensitive promotions. Regularly analyze the effectiveness of these techniques and adjust your approach accordingly to maximize conversions.

4. Encourage user feedback: Actively solicit user feedback through surveys, in-app prompts, or user testing sessions. Utilize this feedback to refine your app's content, features, and user experience, demonstrating to your subscribers that their opinions matter and are instrumental in shaping your app's development.

5. Analyze churn rate and identify areas for improvement: Regularly review your app's churn rate – the percentage of subscribers who cancel their subscription over a given period. Identify common reasons for churn and address these issues to improve your overall retention rate.

6. Segment your subscriber base: Categorize your subscribers based on factors such as subscription tier, usage patterns, and engagement levels. Tailor your marketing and communication

efforts to each segment, ensuring that your messaging is relevant and targeted, ultimately increasing subscriber satisfaction and retention.

By focusing on both dunning management and enhancing subscriber engagement and value, you can create a thriving subscription-based app business. With two decades of experience in software development and entrepreneurship, I can attest that these strategies, when executed effectively, will drive sustainable revenue growth and help you establish a loyal and satisfied subscriber base. Remember that continuous iteration and improvement are crucial to maintaining your app's relevance and value in an ever-evolving app market. So, stay agile, keep learning, and adapt your strategies to achieve lasting success in the world of app monetization.

CHAPTER SEVEN

Data-Driven Monetization: Leveraging Analytics for Success

Key Metrics for Monetization

In this part, we will focus on three critical metrics that are essential for understanding and optimizing your app's monetization strategy. These metrics are Average Revenue Per User (ARPU), Lifetime Value (LTV), and Conversion Rates. By carefully analyzing these metrics, you will be able to make data-driven decisions and fine-tune your monetization strategies to maximize your app's revenue potential.

Average Revenue Per User (ARPU)

The first metric we will discuss is the Average Revenue Per User (ARPU). ARPU is a measure of the total revenue generated by your app divided by the number of active users. This metric is crucial because it provides you with a clear understanding of how much money, on average, you earn from each user. By tracking ARPU over time, you can determine if your app is growing in profitability or if you need to adjust your monetization strategy.

To calculate ARPU, you simply divide your total revenue by the number of active users. For example, if your app generated $5,000 in revenue and had 1,000 active users during the same period, your ARPU would be $5.00.

ARPU is an essential metric to track because it helps you identify trends in user behavior and spending habits. For instance, if you notice a drop in ARPU, it could indicate that users are spending less time in your app, or they are less engaged with the monetization features. By analyzing these trends, you can make strategic decisions to improve user experience, engagement, and ultimately, revenue.

There are several ways to increase your app's ARPU, including:

1. Implementing targeted in-app advertising: By displaying ads that are relevant to your users, you can improve click-through rates and generate more ad revenue.
2. Offering in-app purchases: By providing users with the option to buy additional features, content, or virtual goods, you can generate additional revenue and increase ARPU.
3. Developing a subscription model: By offering users access to premium content or features through a subscription, you can create a steady stream of recurring revenue and increase ARPU.
4. Optimizing pricing strategies: By testing different price points for your app, in-app purchases, or subscriptions, you can identify the most effective pricing strategy to maximize revenue.

Lifetime Value (LTV)

The second key metric we will discuss is the Lifetime Value (LTV) of your users. LTV is the total revenue you can expect to generate from a user during their entire relationship with your app. Understanding LTV is crucial because it helps you determine how much you can spend on acquiring new users while still maintaining a profitable business model.

To calculate LTV, you need to consider several factors, including ARPU, user retention, and the average lifespan of a user in your app. For instance, if your app has an ARPU of $5.00, a user retention rate of 50%, and the average user lifespan is 12 months, your LTV would be:

LTV = ARPU x Retention Rate x User Lifespan
LTV = $5.00 x 0.50 x 12 months
LTV = $30.00

* * *

With this information, you can make informed decisions about your user acquisition strategy. For example, if your LTV is $30.00 and your average cost per user acquisition (CPA) is $10.00, you can confidently invest in user acquisition knowing that you will generate a positive return on investment (ROI).

There are several ways to improve your app's LTV, including:

1. Enhancing user retention: By improving user engagement and satisfaction, you can increase the likelihood that users will continue to use your app over time, resulting in higher LTV.
2. Expanding monetization options: By offering a diverse range of monetization methods, such as in-app purchases, subscriptions, and advertising, you can cater to different user preferences and increase the revenue generated from each user, which ultimately boosts LTV.
3. Focusing on user experience: By optimizing your app's user interface and user experience, you can ensure that users are satisfied with your app and continue to engage with it over time, thereby increasing LTV.
4. Implementing targeted marketing campaigns: By using data-driven marketing techniques to target specific user segments with personalized offers and promotions, you can encourage users to spend more within your app, increasing LTV.

Conversion Rates

The third critical metric we will discuss is Conversion Rates. Conversion rates are a measure of how effectively your app is converting users into paying customers or engaging with monetization features. By tracking conversion rates, you can identify opportunities to optimize your app's monetization strategies and increase overall revenue.

There are several types of conversion rates that you should monitor, including:

1. Free-to-paid user conversion rate: This is the percentage of users who convert from a free user to a paying customer. To

calculate this conversion rate, divide the number of users who made a purchase or subscribed by the total number of users.

2. In-app purchase conversion rate: This is the percentage of users who make an in-app purchase. To calculate this conversion rate, divide the number of users who made an in-app purchase by the total number of users.

3. Ad engagement conversion rate: This is the percentage of users who interact with in-app advertisements, such as clicking on an ad or watching a video ad. To calculate this conversion rate, divide the number of users who engaged with ads by the total number of users.

By monitoring these conversion rates, you can identify trends and opportunities to improve your app's monetization strategy. For example, if you notice a low free-to-paid user conversion rate, you may want to consider offering a limited-time promotion or adjusting your app's pricing strategy to encourage more users to become paying customers.

There are several ways to improve your app's conversion rates, including:

1. A/B testing: By testing different monetization features, pricing strategies, and marketing campaigns, you can identify the most effective methods for converting users into paying customers or engaging with monetization features.

2. Personalizing offers: By using data-driven insights to tailor offers and promotions to specific user segments, you can increase the likelihood that users will engage with your app's monetization features and convert.

3. Optimizing user experience: By making it easy for users to navigate your app and access monetization features, you can encourage more users to engage with these features and improve conversion rates.

In summary, understanding and optimizing key metrics like ARPU, LTV, and Conversion Rates are essential for successfully monetizing your app. By using data-driven insights and continually refining your monetization strategies, you can maximize your app's revenue potential and create a sustainable, profitable business. Remember to always keep an eye on these metrics, as they will guide you in making the right decisions for your app's long-term success.

Analyzing User Behavior and Segmentation

In the modern app economy, data is king. As a software developer and entrepreneur with 20 years of experience in the industry, I can confidently say that leveraging analytics is crucial to understanding your users and maximizing your app's monetization potential. In this subchapter, we will dive into analyzing user behavior and segmentation, focusing on identifying high-value users and customizing offers and promotions for maximum impact.

Identifying High-Value Users

High-value users are the lifeblood of any app, as they tend to be more engaged, loyal, and likely to spend money within your app. Identifying these users allows you to tailor your marketing and monetization strategies to their specific needs and preferences, ultimately driving increased revenue and long-term success.

1. Collect relevant data: To effectively identify high-value users, you must first gather the necessary data. This includes information about user demographics, behavior patterns, in-app purchases, and more. Make sure you have the proper analytics tools in place to track and store this data.

2. Define key performance indicators (KPIs): Establish the KPIs that reflect the characteristics of high-value users. These may include metrics such as revenue per user, session length, retention rate, and conversion rate, among others. KPIs should be specific to your app's goals and business model.

3. Segment your user base: Once you have collected data and defined your KPIs, segment your user base based on their performance against these KPIs. High-value users are those who consistently outperform others in these key areas.

4. Analyze high-value user behavior: Dive deeper into the data to understand the specific behaviors and preferences of your high-value users. What features do they engage with most? When are they most active? How do they respond to

promotions? This information will help you refine your app's monetization strategy to better serve these users.

Customizing Offers and Promotions

Armed with the insights gained from analyzing your high-value users, you can now create targeted offers and promotions designed to increase their engagement and spending. Follow these steps to effectively customize offers and promotions for your most valuable users:

1. Develop user personas: Create detailed profiles of your high-value users, including demographics, preferences, motivations, and pain points. This will help you better understand their needs and tailor your offers accordingly.
2. Create targeted offers: Design special offers and promotions that appeal specifically to your high-value users. This could include exclusive content, discounts on in-app purchases, or early access to new features.
3. Personalize messaging: When communicating offers and promotions to your high-value users, be sure to personalize your messaging. Use their name, reference their past behavior, and highlight the benefits that are most relevant to them. This will help increase the likelihood that they will engage with your offer and make a purchase.
4. Test and optimize: Continuously test and optimize your offers and promotions to ensure maximum effectiveness. Experiment with different messaging, formats, and targeting to determine what resonates best with your high-value users.
5. Track results: Monitor the performance of your offers and promotions by tracking key metrics such as conversion rate, revenue per user, and customer lifetime value. Use this data to refine your strategy and further improve your targeting and personalization efforts.
6. Expand targeting: Once you have successfully engaged your high-value users, consider expanding your targeting efforts to other user segments with similar characteristics. This will help you broaden your reach and drive additional revenue.

By implementing a data-driven approach to monetization, you can identify your high-value users and customize offers and promotions to meet their specific needs and preferences. This will not only lead to increased revenue and user satisfaction but also help you create a more sustainable and profitable business.

1. Leverage machine learning: As your app grows and your user base expands, it becomes increasingly challenging to manually analyze and segment users. By leveraging machine learning algorithms, you can automate the process of identifying high-value users and predicting their future behavior. This allows you to scale your monetization efforts and continually refine your targeting and personalization strategies.

2. Utilize predictive analytics: By analyzing historical data and identifying patterns, you can use predictive analytics to forecast user behavior and anticipate future needs. This enables you to proactively address potential issues and capitalize on emerging opportunities, ensuring your app remains relevant and engaging for your high-value users.

3. Foster user loyalty: High-value users are not only more likely to spend money within your app, but they are also more likely to recommend it to others. Encourage user loyalty by consistently delivering value, soliciting feedback, and incorporating user suggestions into future updates and improvements.

4. Monitor trends and industry developments: Stay informed about emerging trends and developments within the app industry to ensure your monetization strategies remain current and effective. This includes keeping an eye on your competitors and learning from their successes and failures, as well as staying up-to-date on the latest technologies, platforms, and advertising opportunities.

In conclusion, adopting a data-driven approach to app monetization is essential for understanding user behavior, identifying high-value users, and crafting targeted offers and promotions. By leveraging analytics, machine learning, and predictive technologies, you can create a more personalized and engaging experience for your users, ultimately driving increased revenue and long-term success. As an experienced software developer and entrepreneur, I can attest to the power of data and its ability to transform an app idea into a thriving and profitable business.

A/B Testing and Optimization

As an app developer and entrepreneur with 20 years of experience, I understand the critical role that data-driven strategies play in the monetization of an app. One such approach that has proven immensely valuable is A/B testing and optimization. By continuously testing, analyzing, and refining various aspects of your app, you can make informed decisions that will drive user engagement, satisfaction, and ultimately, revenue. In this subchapter, we will explore how to effectively experiment with pricing and offers, as well as refine ad placements and formats, to maximize your app's earning potential.

Experimenting with Pricing and Offers

1. Establish clear objectives: Before conducting any A/B testing, it's essential to define your goals and objectives clearly. Are you looking to increase revenue, boost user engagement, or improve user retention? Having a clear understanding of your desired outcomes will help guide your testing and ensure your efforts are focused and effective.

2. Test various pricing models: One of the most critical factors influencing user spending is the pricing of your in-app purchases or subscriptions. Experiment with different pricing models, such as tiered pricing, pay-per-use, or flat-rate subscriptions, to determine which approach resonates best with your target audience.

3. Offer limited-time promotions: Limited-time promotions can create a sense of urgency among your users, motivating them to make a purchase or upgrade. Test various promotional strategies, such as flash sales, holiday discounts, or seasonal offers, to identify which promotions yield the best results.

4. Experiment with bundle deals: Offering multiple items or features as part of a discounted bundle can encourage users to spend more within your app. Test various bundle combinations and pricing to identify the most attractive and profitable deals for your users.

5. Test pricing elasticity: Understand the price sensitivity of your

users by testing various price points for your in-app purchases or subscriptions. This will help you identify the optimal pricing strategy that maximizes revenue without alienating potential customers.

Refining Ad Placements and Formats

1. Choose the right ad formats: Different ad formats, such as banner ads, interstitial ads, and rewarded ads, may perform differently depending on your app and audience. Experiment with various ad formats to identify which types generate the most revenue and provide the best user experience.

2. Test ad placements: The location of ads within your app can significantly impact user engagement and revenue generation. Conduct A/B tests to determine the most effective ad placements that capture user attention without disrupting their experience.

3. Optimize ad frequency and timing: Displaying too many ads too frequently can lead to user frustration and app abandonment. Test different ad frequencies and display times to strike the right balance between generating revenue and maintaining a positive user experience.

4. Analyze user engagement metrics: Review metrics such as click-through rates, conversion rates, and ad revenue to assess the effectiveness of your current ad placements and formats. Use these insights to guide your A/B testing and optimization efforts.

5. Collaborate with ad networks: Work closely with your chosen ad networks to gain insights into industry trends, best practices, and new ad formats. Leverage this knowledge to inform your testing and optimization efforts, ensuring your app remains competitive in the rapidly evolving ad landscape.

6. Continuously iterate and improve: A/B testing and optimization is an ongoing process. Continuously analyze your data, make adjustments, and test new ideas to ensure your app's monetization strategy remains effective and profitable.

In summary, effective A/B testing and optimization can help you identify the most successful pricing strategies, promotional offers, ad placements, and formats for your app. By continuously experimenting and refining these elements, you can make data-driven decisions that will maximize revenue, enhance user experience, and contribute to the overall success of your app. Remember, the key to successful A/B testing and optimization is to approach it as an ongoing, iterative process rather than a one-time event.

As you progress through the various stages of testing and optimization, keep these additional tips and best practices in mind:

1. Be patient and persistent: A/B testing requires time and patience. It's essential to give each test enough time to generate meaningful results. Don't rush to conclusions or make hasty decisions based on limited data.
2. Focus on statistically significant results: Ensure that your tests generate statistically significant results before making any decisions. This will help you avoid making changes based on random fluctuations or anomalies in your data.
3. Prioritize high-impact changes: Focus on testing and optimizing elements that have the potential to make a significant impact on your app's revenue generation and user experience. This will help you maximize the return on your testing efforts.
4. Encourage a culture of experimentation: Foster a culture within your development team that encourages continuous experimentation and learning. This mindset will help drive innovation and keep your app ahead of the competition.
5. Monitor the competition: Stay informed about the latest trends, best practices, and strategies employed by your competitors. Use this knowledge to inform your A/B testing and optimization efforts and ensure your app remains competitive in the market.
6. Measure the impact of changes: After implementing changes based on your A/B testing results, continue to monitor your app's performance to assess the impact of those changes. This will help you refine your optimization efforts further and ensure that you're making data-driven decisions that contribute to your app's ongoing success.

By following these guidelines and continuously iterating on your app's monetization strategy through A/B testing and optimization, you will be well-equipped to maximize revenue generation, enhance user experience, and build a sustainable and profitable app business.

In conclusion, A/B testing and optimization play a crucial role in the success of any app monetization strategy. By experimenting with pricing, offers, ad placements, and formats, you can make data-driven decisions that will not only increase revenue but also improve the overall user experience.

Remember, the key to successful A/B testing is to approach it as an ongoing, iterative process that requires patience, persistence, and a commitment to continuous improvement. With the right approach and mindset, you can leverage A/B testing and optimization to build a thriving app business and achieve long-term success in the competitive world of app monetization.

CHAPTER EIGHT

Scaling Your App: Expansion and Growth Strategies

Internationalization and Localization

In today's globalized and interconnected world, expanding your app's reach to international markets is more important than ever. Tapping into new markets can significantly increase your user base, revenue, and overall success. This chapter will guide you through the process of internationalization and localization, with a focus on adapting your app for global markets and tailoring regional pricing and offers.

Adapting Your App for Global Markets

The first step in internationalizing your app is to ensure it is technically and culturally adapted for different markets. This process involves making your app compatible with various languages, currencies, and devices, as well as considering cultural sensitivities and regional preferences.

Language Support: Providing language support for your target markets is critical to attracting and retaining users. Utilize professional translation services or work with native speakers to translate your app's content, metadata, and marketing materials. This will not only improve user experience but also boost your app's visibility in local app stores.

Currency Support: Ensure that your app supports local currencies for in-app purchases, subscriptions, and ads. This will make it easier for users to understand and evaluate the cost of your app and its offerings. Using a payment processing platform that supports multiple currencies can simplify this process and reduce friction for users.

Device Compatibility: As you expand into new markets, consider the devices and operating systems used by your target audience. Adapt your app to be compatible with the most popular devices and platforms in each region, ensuring that your app runs smoothly and provides an optimal user experience.

Cultural Sensitivities: Be mindful of cultural differences when adapting your app for international markets. Avoid using images, icons, or language that could be offensive or misunderstood in different cultures. Research local customs, traditions, and beliefs to ensure that your app is culturally appropriate and respectful.

Regional Preferences: Customize your app's design, features, and content to align with regional preferences and trends. For example, you may need to modify your app's color scheme to align with local aesthetic preferences or adapt your app's user interface to accommodate language-specific text direction.

Regional Pricing and Offers

Once you have adapted your app for global markets, it's time to tailor your pricing strategy and offers to different regions. This will ensure that your app remains competitive and appeals to local users.

Market Research: Conduct thorough market research to understand the competitive landscape and pricing strategies in each target market. This will help you determine the appropriate price points and offers for your app in each region. Make sure to consider factors such as local purchasing power, currency conversion rates, and regional economic conditions.

Price Localization: Localize your app's pricing to make it more appealing and accessible to users in different regions. This may involve adjusting the base price, offering regional discounts, or adopting tiered pricing structures based on local income levels. By pricing your app appropriately for each market, you can maximize revenue and ensure that your app is accessible to a broader audience.

Regional Offers and Promotions: Tailor your in-app offers and promotions to cater to regional preferences and events. This may involve creating region-specific content, features, or virtual goods, or offering discounts and promotions tied to local holidays and celebrations. By providing relevant and appealing offers, you can drive user engagement and revenue in each market.

Payment Methods: Support locally preferred payment methods to make it easy for users in different regions to purchase your app and its offerings. This may involve integrating with local payment gateways or offering alternative payment options, such as carrier billing or mobile wallets.

Monitor Performance and Adjust: Continuously monitor the performance of your app in each market and use data-driven insights to refine your pricing strategy and offers. This may involve A/B testing different price points or promotions, adjusting regional discounts, or iterating on localized content and features. By staying agile and responsive to market conditions, you can optimize your app's monetization strategy for each region.

Localization of Marketing Efforts: Ensure that your marketing and advertising campaigns are tailored to each target market. This includes localizing ad creatives, promotional materials, and messaging to resonate with users in different regions. Also, consider partnering with local influencers, publishers, or media outlets to increase your app's visibility and credibility in each market.

Legal Compliance: When expanding into new markets, it's crucial to comply with local laws and regulations, such as data protection, consumer protection, and tax laws. Consult with local legal experts to ensure that your app, its features, and monetization strategies are compliant with the laws in each target market.

App Store Optimization (ASO): Optimize your app store listing for each target market to increase visibility and downloads. This involves localizing your app's title, description, keywords, and screenshots to appeal to users in each region. Regularly monitor and update your app store listing to stay competitive and responsive to changing market conditions.

Customer Support: Provide localized customer support to users in different regions, ensuring that they can access help and assistance in their native language. This may involve hiring multilingual support staff, partnering with local support agencies, or leveraging translation tools and services.

Ongoing Localization and Optimization: Expanding your app to international markets is an ongoing process that requires continuous optimization and iteration. Regularly review and update your app's localization, pricing, offers, and marketing efforts to ensure that they remain relevant, competitive, and appealing to users in each region.

In conclusion, successfully scaling your app to international markets involves careful planning, research, and execution. By adapting your app for global markets, tailoring regional pricing and offers, and optimizing your app's monetization strategy based on data-driven insights, you can significantly increase your app's reach, revenue, and overall success.

By implementing the strategies outlined in this chapter, you can effectively expand your app's presence in international markets and create a truly global app that appeals to users worldwide. Remember that the key to success lies in staying agile, responsive, and data-driven, and always keeping the needs and preferences of your users at the heart of your app's development and monetization strategies.

Cross-Platform Development

As your app gains traction and user base, it is essential to consider expanding to multiple platforms to maximize your app's reach and revenue potential. In this chapter, we will discuss cross-platform development, focusing on expanding to iOS, Android, and beyond, as well as platform-specific monetization tactics that can be employed to optimize your app's revenue generation across various platforms.

Expanding to iOS, Android, and Beyond

Understanding Platform Differences: Before diving into cross-platform development, it's crucial to understand the differences between iOS, Android, and other platforms. These differences include design guidelines, user demographics, and market share. For example, iOS users are typically known for higher average spending, while Android has a more extensive global market share. Research each platform's user base and demographics to identify which platforms are best suited for your app.

Choosing a Cross-Platform Development Approach: There are several approaches to cross-platform development, including native development, hybrid development, and cross-platform development frameworks. Native development involves building separate apps for each platform using platform-specific programming languages and tools, while hybrid development involves building an app with a single codebase that runs on multiple platforms. Cross-platform development frameworks, such as Flutter, React Native and Xamarin, allow developers to write code once and deploy it across multiple platforms. Choose the development approach that best aligns with your app's goals, requirements, and resources.

Designing for Platform-Specific UX: When expanding to multiple platforms, ensure that your app adheres to each platform's design guidelines and best practices. iOS and Android have distinct design languages and user interface conventions, such as navigation patterns and button styles. By designing your app to feel native on each platform, you can create a more intuitive and enjoyable user experience that drives user engagement and retention.

Managing Platform-Specific Features and Integrations: Each platform offers unique features and services, such as Apple's App Store Connect, Google's Firebase, or Microsoft's Azure. Evaluate these services and integrate them into your app to take advantage of platform-specific features, such as in-app purchases, push notifications, and analytics. Additionally, be mindful of platform-specific restrictions, such as iOS's strict app review process and Android's varying device capabilities.

Platform-Specific Monetization Tactics

Adapting Monetization Strategies for Each Platform: Due to differences in user behavior and spending patterns between platforms, it's essential to adapt your monetization strategies accordingly. For example, iOS users may be more inclined to make in-app purchases, while Android users may respond better to ads. Analyze user data and market research to determine the most effective monetization strategies for each platform.

Platform-Specific Ad Networks: When implementing in-app advertising, consider using platform-specific ad networks, such as iAd for iOS or AdMob for Android. These ad networks are designed to work seamlessly with their respective platforms and may offer better performance, targeting, and fill rates compared to generic ad networks.

Platform-Specific Pricing: Different platforms may have varying user demographics and spending patterns, so it's essential to tailor your app's pricing accordingly. For example, you may choose to offer a lower-priced subscription tier for Android users, given the platform's generally lower average spending. Alternatively, you may decide to offer different pricing structures for different regions to account for varying purchasing power.

Platform-Specific Promotions and Offers: To increase user acquisition and monetization, consider offering platform-specific promotions and offers, such as limited-time discounts, exclusive in-app purchases, or platform-specific bonuses. By tailoring your promotions to each platform's user base, you can maximize the effectiveness of your marketing campaigns and increase user spending.

Platform-Specific Analytics and Optimization: Analyze platform-specific user data to identify trends, behaviors, and areas for improvement. Use this information to refine your monetization strategies and user experience for each platform. For example, if you find that iOS users are more likely to make in-app purchases, consider focusing on optimizing the in-app purchase experience for that platform. Similarly, if Android users are more receptive to in-app advertising, refine your ad placements and formats to maximize revenue from that platform.

* * *

Platform-Specific User Acquisition: To attract users to your app on each platform, develop targeted user acquisition strategies. For example, leverage platform-specific advertising networks, such as Apple Search Ads or Google Ads, to reach users who are actively searching for apps like yours. Additionally, optimize your app store listings and metadata for each platform to improve organic visibility and conversion rates.

Leveraging Platform-Specific Partnerships: Building relationships with platform owners, such as Apple or Google, can help your app gain visibility and credibility. These relationships can lead to featured placements in app stores, co-marketing opportunities, or even early access to new platform features. Attend platform-specific developer events, such as Apple's WWDC or Google I/O, to network with platform representatives and stay up to date on the latest platform advancements.

Continuous Cross-Platform Development: As you expand to multiple platforms, it's essential to continuously iterate and improve your app on each platform. Keep abreast of platform updates, new features, and emerging trends to ensure that your app remains competitive and relevant in the ever-changing app landscape. Regularly evaluate your app's performance, user feedback, and market trends to inform your cross-platform development roadmap.

In conclusion, scaling your app across multiple platforms can significantly increase your app's reach, user base, and revenue potential. To successfully navigate the complexities of cross-platform development, it's crucial to understand each platform's unique characteristics and adapt your monetization strategies, user experience, and development processes accordingly.

By leveraging platform-specific tactics and continuously refining your app based on data-driven insights, you can maximize your app's success across iOS, Android, and beyond.

Mergers, Acquisitions, and Partnerships

Identifying Strategic Opportunities

1. Analyze the Competitive Landscape: To identify potential merger, acquisition, or partnership opportunities, start by thoroughly analyzing the competitive landscape of your app's niche or industry. This includes assessing the strengths and weaknesses of other apps in your market and understanding their unique selling points. Look for gaps in the market, where your app could benefit from partnering with or acquiring complementary products or services. Consider both direct competitors and adjacent companies that target similar user bases or share a common value proposition.

2. Set Clear Goals: Before exploring potential opportunities, set clear objectives for your desired outcomes. These goals might include expanding your user base, increasing revenue, entering new markets, acquiring new technologies or expertise, or streamlining operations. By establishing specific goals, you can more effectively evaluate potential opportunities and ensure that any potential deals align with your overall business strategy.

3. Evaluate Potential Synergies: When considering potential partners, mergers, or acquisitions, evaluate the synergies between your app and the target company. Synergies may arise from shared user bases, complementary features or services, similar marketing strategies, or operational efficiencies. Assess the potential value that can be created through collaboration or integration, and weigh this against the risks and costs associated with the deal.

4. Financial Viability: Ensure that any potential merger, acquisition, or partnership is financially viable and aligns with your company's financial goals. This involves evaluating the target company's financial performance, growth potential, and the potential return on investment. Consider the financial resources required to execute the deal, and whether your company has the necessary capital or financing to support the transaction.

Collaborating for Mutual Success

1. Establishing Clear Communication Channels: Effective communication is critical to the success of any partnership,

merger, or acquisition. Establish clear communication channels between all parties involved, including regular meetings, shared project management tools, and transparent reporting structures. Ensure that all stakeholders are kept informed of progress, challenges, and key decisions throughout the collaboration process.

2. Defining Roles and Responsibilities: To ensure a smooth collaboration, clearly define the roles and responsibilities of each party involved. This should include outlining expectations for each team, identifying key decision-makers, and establishing processes for resolving disputes or disagreements. By establishing clear roles and responsibilities upfront, you can minimize potential conflicts and ensure that all parties are aligned towards a common goal.

3. Integration Planning: In the case of mergers and acquisitions, thorough integration planning is essential to realizing the full value of the deal. This involves assessing the compatibility of each company's operations, systems, and processes, and developing a detailed plan for integrating these elements. Consider the potential impact on your existing users and develop strategies for minimizing disruption or negative effects on the user experience.

4. Fostering a Collaborative Culture: To ensure the long-term success of any partnership, merger, or acquisition, it's essential to foster a collaborative culture between the involved parties. This includes promoting open communication, encouraging innovation and idea-sharing, and celebrating shared successes. By cultivating a positive and collaborative working environment, you can increase the likelihood of achieving your desired outcomes and creating lasting value for all stakeholders.

5. Continuous Improvement and Evaluation: As with any aspect of your app business, it's important to continuously evaluate the success of your partnerships, mergers, or acquisitions, and identify areas for improvement. Regularly review key performance indicators, such as user growth, revenue, and user satisfaction, to assess the impact of your collaboration. Iterate and refine your approach based on these insights to maximize the value of your partnerships over time.

* * *

In summary, identifying strategic opportunities for mergers, acquisitions, and partnerships can play a crucial role in the expansion and growth of your app business. By carefully evaluating potential synergies, setting clear goals, and fostering a collaborative culture, you can successfully collaborate with other companies to achieve mutual success and take your app business to new heights. Remember that the key to successful collaboration lies in open communication, strategic planning, and a shared vision for growth.

1. Leveraging Networks and Industry Connections: Identifying strategic opportunities often involves leveraging your existing networks and industry connections. Attend industry conferences, events, and networking sessions to build relationships with potential partners or acquisition targets. Engage with industry influencers, thought leaders, and other stakeholders who can provide valuable insights and connections to help you identify potential collaboration opportunities.

2. Legal and Regulatory Considerations: When exploring potential mergers, acquisitions, or partnerships, it's essential to consider legal and regulatory implications. Consult with legal counsel to ensure that any potential deals comply with relevant laws and regulations, such as antitrust or data privacy requirements. Additionally, consider the potential impact of regulatory changes on your partnership or acquisition strategy, and plan for potential risks accordingly.

3. Negotiating and Structuring the Deal: Successful mergers, acquisitions, and partnerships often involve complex negotiations and deal structuring. Work with experienced advisors or consultants to help you navigate these processes, and ensure that the terms of the deal align with your strategic goals and financial objectives. When negotiating the deal, be prepared to make compromises, but also know your limits and be willing to walk away if the terms do not meet your expectations.

4. Post-Deal Integration and Support: After completing a merger, acquisition, or partnership, focus on successfully integrating the new company or partner into your existing operations. Provide ongoing support and resources to ensure a smooth transition, including training, technical assistance, and access

to necessary tools or systems. Continuously monitor and evaluate the integration process, addressing any challenges or issues that arise to ensure the long-term success of the collaboration.

5.   Learning from Experience and Adapting: As you embark on mergers, acquisitions, or partnerships, it's essential to learn from your experiences and adapt your approach over time. Reflect on the successes and challenges of past collaborations, and apply these insights to future opportunities. Continuously refine your strategy and approach to maximize the value of your partnerships and drive ongoing growth for your app business.

In conclusion, mergers, acquisitions, and partnerships can be powerful growth strategies for app developers and entrepreneurs. By identifying strategic opportunities, carefully evaluating potential synergies, and fostering collaborative relationships, you can leverage these deals to expand your user base, increase revenue, and achieve long-term success.

Remember to continuously evaluate and refine your approach, learning from your experiences and adapting to the ever-evolving app market landscape.

CHAPTER NINE

Conclusion

Your Journey to App Monetization Mastery

As we reach the end of this comprehensive guide, it is important to reflect on the key points and lessons learned throughout the chapters. The road to app monetization mastery is paved with numerous decisions, strategies, and experiments, and by now, you should have a solid understanding of the various approaches and techniques that successful app developers and entrepreneurs have used to turn their app ideas into profitable businesses.

One of the most critical aspects of app monetization is understanding your target audience and their preferences. By conducting thorough market research and user testing, you can gain valuable insights into your users' needs and desires, which will inform your app's design, features, and monetization methods. Remember to always put your users first and strive to create a seamless, enjoyable user experience that keeps them engaged and encourages them to spend within your app.

We discussed various monetization strategies such as in-app purchases, subscription models, advertising, and data-driven monetization, each with its own set of advantages and challenges. It is crucial to carefully consider which methods are best suited for your app and target audience, and to remain flexible and adaptable as your app evolves and grows. Continuously test, iterate, and optimize your monetization strategies to maximize revenue while maintaining user satisfaction.

User retention and engagement are paramount to your app's long-term success. Implementing loyalty programs, rewards, push notifications, and in-app messaging can help keep your users engaged and coming back for more. In addition, analyzing user behavior and segmentation allows you to identify high-value users and customize offers and promotions to cater to their preferences, further enhancing their experience and increasing their likelihood to spend.

Embrace the power of A/B testing and optimization to refine every aspect of your app monetization strategy, from pricing and offers to ad placements and formats. By consistently testing and learning from the results, you can make data-driven decisions that propel your app to new heights.

Expanding your app's reach is another crucial component of scaling your app and increasing revenue. Focus on internationalization and localization to make your app accessible and appealing to users across the globe, and consider expanding to multiple platforms such as iOS, Android, and beyond to tap into new markets and audiences. Be mindful of regional pricing and offers to ensure that your app remains competitive and appealing to users in different countries and regions.

Forming strategic partnerships, mergers, or acquisitions can also bolster your app's growth and revenue potential. By identifying synergistic opportunities and collaborating with other companies or developers, you can benefit from shared resources, knowledge, and user bases, further accelerating your app's success.

Finally, it is important to recognize that app monetization is an ongoing journey that requires perseverance, creativity, and adaptability. The app market is constantly evolving, and staying ahead of the curve requires continuous learning, experimentation, and improvement. Keep an eye on emerging trends, technologies, and best practices, and be prepared to pivot or adapt your strategies as necessary.

In conclusion, mastering app monetization is a complex and multifaceted endeavor, but with the knowledge and insights gained from this guide, you are well-equipped to navigate the challenges and opportunities that lie ahead. By applying the strategies and techniques discussed throughout these chapters, you will be well on your way to turning your app idea into a thriving, profitable business. So, embark on your journey to app monetization mastery with confidence and determination, and may your app achieve the success it deserves.

As you progress on your app monetization journey, remember that the lessons and strategies outlined in this guide are not meant to be followed dogmatically but rather serve as a foundation upon which you can build your own unique approach. The most successful app developers and entrepreneurs are those who remain open-minded, agile, and innovative, constantly pushing the boundaries and challenging conventional wisdom in search of new, more effective ways to monetize their apps.

Keep in mind that the app industry is ever-changing, with new platforms, technologies, and user trends emerging regularly. Staying informed and up-to-date on the latest developments in the app ecosystem is essential to maintaining a competitive edge and ensuring that your monetization strategies remain effective and relevant. Participate in industry events, engage with fellow developers and entrepreneurs, and actively seek out new ideas and insights that can help you grow your app business.

Moreover, don't be afraid to take risks and experiment with unconventional monetization methods or strategies. While some experiments may fail, others may yield unexpected and significant rewards. By embracing a culture of experimentation and learning from both your successes and failures, you can iterate and improve your app's monetization strategy more effectively than if you were to strictly adhere to established best practices.

In addition, cultivate a strong network of peers, mentors, and advisors who can provide valuable guidance, feedback, and support throughout your app monetization journey. The collective wisdom and experience of others can be an invaluable resource, helping you to avoid common pitfalls and discover new opportunities that you may not have identified on your own.

* * *

Lastly, always maintain a user-centric mindset and strive to create an app experience that delights, engages, and provides genuine value to your users. At the end of the day, your app's success hinges on its ability to attract, retain, and monetize users. By prioritizing user satisfaction and delivering a product that meets their needs and desires, you can build a loyal user base that will not only drive revenue but also serve as ambassadors for your app, helping to fuel its growth through word-of-mouth referrals and positive reviews.

In closing, I hope that this guide has provided you with the tools, knowledge, and inspiration needed to embark on your journey to app monetization mastery with confidence and determination. By applying the principles and strategies discussed in these chapters, you can transform your app idea into a thriving, profitable business that stands the test of time. Remember, the road to app monetization mastery is a continuous journey of learning, experimenting, and adapting, so stay curious, stay persistent, and above all, stay focused on creating an exceptional app experience that your users will love. Good luck, and here's to your app's success!